RECOVERING RESOURCES
– RECYCLING CITIZENSHIP

Recovering Resources
– Recycling Citizenship
Urban Poverty Reduction in Latin America

JUTTA GUTBERLET
University of Victoria, Canada

Routledge
Taylor & Francis Group

LONDON AND NEW YORK

First published 2008 by Ashgate Publishing

2 Park Square, Milton Park, Abingdon, Oxon OX14 4RN
711 Third Avenue, New York, NY 10017, USA

Routledge is an imprint of the Taylor & Francis Group, an informa business

First issued in paperback 2016

British Library Cataloguing in Publication Data
Gutberlet, Jutta
    Recovering resources - recycling citizenship : urban
    poverty reduction in Latin America
    1. Recycling (Waste, etc.) - Latin America 2. Sociology,
    Urban - Latin America 3. Poverty - Government policy -
    Latin America
    I. Title
    363.7'282'098

Library of Congress Control Number: 2008920975

ISBN 13: 978-0-7546-7219-7 (hbk)
ISBN 13: 978-1-138-26622-3 (pbk)

Contents

List of Figures

List of Illustrations

List of Tables

Acknowledgements

A strong desire to contribute to the debate on community oriented, participatory and appropriate local development is the central motivation for this book. Resource intense and wasteful lifestyles, perverse habits of over-consumption and widespread prevailing inadequate waste management need to be addressed, not only with critique but with innovative resolutions. The pressing demands for social justice and more environmental health are visible locally and obviously affect the global scale. This is becoming evident with climatic change, loss of biodiversity, increase in violence and a rapidly progressing human-nature disconnect. Reflecting on the various facets linked to production, consumption, disposal and recovery elucidates the fundamental ecological and social condition of interconnectedness, illustrating possible limitations of our development model and highlighting the deep ecological and human geographic impacts. The current book brings together findings from research conducted in Brazil and is based on ideas and thoughts from many readings and lived experiences particularly from a Latin American perspective. The geographers Aziz Nacib Ab'Saber, Milton Santos, José Pereira de Queiroz Neto, Gerd Kohlhepp and the educator Paulo Freire have certainly inspired me, particularly in adopting an integrated interpretation of human and natural facts in space and over time.

My deep appreciation goes to the *catadores* and *catadoras* (recyclers), the true heroes that in extreme poverty and exclusion recover in their daily activity the wasted resources, and thus contribute to a cleaner environment. Without their inspiration and persistence, their spirit and local knowledge this book would not have been written. The many encounters in praxis and theory with my dear friends and colleagues in Brazil: Ruth Takahashi, Fábio Cardozo, Angela Baeder, Ana Maria Marins and Lúcia Campos have preciously enriched my learning and thinking. I am very grateful for their insights.

The magic and enchanting historical atmosphere of the city of Venice, where I spend part of my sabbatical semester in 2007 and the infrastructure offered by the University *Ca'Foscari* were essential for the dream of writing this book to become true. I deeply enjoyed the pleasure of diving into the literature, making sense of innovative thoughts and translating my reflections into writing. In Canada, my friends Thomas Heyd and Rick Searl have given me the necessary initial encouragement to write this book for a wider public and their feedback has stimulated persistance during difficult times. Thanks to the many discussions with my students, in particular Crystal Tremblay, examining consumption and waste issues in the light of urban sustainability I could find the inspiration for this endeavour.

My parents have given me the formal and informal education providing the indispensable, life-embracing attitudes and tools that ultimately have enabled me to crystallize my dedication and to develop and follow this particular path.

The unconditional moral support and encouragement of my friends in Canada (in particular from the *Tertulia*), in Europe and in Brazil have made this process a liveable and worthwhile one. I am also appreciative for the opportunities granted by the University of Victoria to facilitate my research, teaching and lifelong learning. I am very grateful for the contribution from Ole J. Heggen, the cartographer of the Department of Geography at the University of Victoria, in producing the maps and graphics for this book.

Finally, I would like to recognize: the community *Pedra sobre Pedra* in São Paulo and specifically the members of the local neighbourhood associations, the *catadores* from the *Vida Limpa* programme in Diadema, *Cooperpires* in Ribeirão Pires, and all the leaders of the *Fórum Recicla São Paulo*, the *Participatory Sustainable Waste Management* Project (*Projeto Brasil-Canadá*) and the *Rede Gerando Renda* project.

Chapter 1

Waste no Waste

Introduction: The North – South Divide of One World

This work on urban recycling and poverty reduction is the result of the compilation of many years of research and lived experiences with social development issues in Latin America, particularly in Brazil. It combines perspectives from the South – the place of my heart – and from the North – the place of my roots. Growing up in Brazil and having lived, studied and worked in different places in this country for many years has provided me with insights into the society, its culture, history and economy, establishing a deep connection with the people and the land. I have travelled the country extensively and have learned about the beauty and diversity of the population and the landscapes. These experiences, I believe, have been crucial to my understanding of the often life-threatening survival conditions of the poor. It is important to know about the multifarious perspectives of the country in order to be able to search for alternatives and resolutions to these predicaments. Several opportunities to be in different places in Africa, Asia and Europe have also shaped my perception, knowledge and rationale. Waste and poverty, unfortunately, are ubiquitous problems that need to be addressed in the current sustainability debate.

The book is about the critical environmental and social challenges that come together with consumption and the generation of waste. It takes place in the urban landscapes of poverty in the developing world. However, the writing is about the hope of the people who, in reality, are the true heroes in a society caught up in over-consumption and disposable lifestyles. It is about the *binners*, *catadores*, *carrinheiros*, *cartoneros*, *recuperadores*, *zabaleen* or *mikhalas*, as the recyclers are called in different cultural contexts. I discuss the complex socio-political and environmental facets of solid waste and its management and touch on the wider implications of production and consumption, raising questions about and alternatives to the environmental and social facets of waste.

Do We Need to Change our Concept of Waste?

The way we understand and define waste has profound implications on how we deal with it on a daily basis. From a resource conservation perspective, there is no such thing as waste, and dumping corresponds to wasting resources (Ackerman and Mirza 2001; Drackner 2005; Pongracz and Pohjola 2004). This perspective underlines a pro-active attitude of thinking about waste disposal before production and consumption. From a resource conservation perspective, the producer is accountable for the waste of resources during production and for the consequences of disposal.

The consumer is responsible for the diversion and reintroduction of waste into the correct resource flows. When we throw away something, we also throw away all the embodied energy that was used to produce the item – a loss that is not accounted for by traditional economic thinking. It is important to close the loop from production to consumption and back. Ultimately the act of not wasting diminishes the risks of resource depletion, keeps the environment clean and thus contributes to sustainability. Voluntary simplicity is 'a generic term for variously motivated contemporary phenomenon: the foregoing of maximum consumption and, possibly, income' (Shaw and Newholm 2002, 169). It is evident that avoiding the generation of any waste, through ethical responsible consumption (Friedman 1996) or voluntary simplicity and downshifting (Etzioni, 1998) is imperative in addressing the pressing current environmental and social predicaments. There is a wide range of different practices related to less wasteful and more aware consumption. Ernst Friedrich Schumacher was right with the statement *Small is beautiful*; this also should be reflected in our consumption patterns and lifestyles (Schumacher 1999).

Similar to the *polluter pays* principle, where the polluting party pays for the damage done to the natural environment (also known as Extended Polluter Responsibility); *producer responsibility* is a policy tool that aims to ensure that businesses take responsibility for products they sell once those products have reached the end of their lives (*cradle to cradle principle*) (Robins and Kumar 1992). In many countries of the European Union, this principle is applied to packaging waste, like the *green dot* programme (*Grüner Punkt*) in Germany, where the Duales System Deutschland GmbH has to take responsibility for collection of all packaging marked with a green dot and arranges either for take-back according to pre-established liabilities with the manufacturers or eco-friendly recovery. It offers the manufacturers and distributors of sales packaging a comprehensive service package. Nevertheless, it is often unclear whether the materials are really reintroduced into production or whether they are incinerated or even exported to locations with cheaper labour. In theory the application of the *producer responsibility* concept to all stages of the product lifecycle can solve and prevent many environmental problems (Sheehan and Spiegelman 2005).

Proactive and preventive alternatives are needed. Hartwick (2000) outlines effective ways to minimize waste. The concepts *food mileage, ecological footprint and commodity chains* help identify and understand the linkages between production and consumption. Labelling is important to support consumer decisions; however, it is unreasonable to expect that consumers always read the instructions on all purchased products. Information campaigns are fundamental to create socially and environmentally responsible consumption (Shaw and Newholm 2002). Reusing products or packaging, composting, donating and purchasing reusable materials are forms of active waste minimization behaviour, which are becoming more acceptable. Changing lifestyles and consumption patterns may ultimately contribute to sustainability. Consumption trends that minimize the use of natural resources, the generation of toxic materials, emissions of waste and pollutants, over the lifecycle of the product, as suggested by the *Brundtland Report*, need to gain momentum (see also Denison 1996).

The term sustainability has been widely debated. It…'can mean any important changes in values, public policies and public and private activity that moves communities and individuals towards realization of the key tenants of ecological integrity, social harmony, and political participation' (Weber 2003, 194). Here the environmental and temporal dimensions are recognized. With a focus on sustainable communities the approach becomes more tangible, addressing the day-to-day lives of people and the emerging negative impacts as a consequence of their lifestyle. Community is, according to Maser (1997) '…a group of people with similar interests…in a shared locality…and…a common attachment to their place of residence where they have some degree of local autonomy' (99). Hempel (1999, 48) defines sustainable communities as 'communities in which economic vitality, ecological integrity, civic democracy, and social wellbeing are linked in complementary fashion, thereby fostering a high quality of life and a strong sense of reciprocal obligations among its members' (cited in Weber 2003, 194). To build a sustainable community is a visionary project, a continuous and dynamic process.

A shift in values, considering waste as a resource not to be wasted, is part of the paradigm shift towards more sustainable societies. The official definitions from the EU, the OECD or UNEP illustrate the common view of waste as something that is unwanted or that its holder has disposed of or is going to discard (Pongracz and Pohjola 2004). Waste is still primarily seen as a nuisance, as something the owner doesn't want to have, or something that is abandoned. Related to this view of waste is the widespread negative perception of those who work with waste, including informal recyclers. These people are often associated with trash and doing the dirty work, and they are stigmatized by mainstream society as a result.

Sociological and anthropological reflections as well as ethnographic studies contribute to a better understanding of how we interpret waste and how we can stimulate resource conservation consciousness among consumers. We need to consider how we can practice social justice with the most excluded and impoverished segment of society that is involved in recovering recyclables. The literature from different disciplinary backgrounds and experiences from various countries contribute a knowledge base that is useful in the transition towards sustainable societies, embracing change in our consumption-oriented culture, recognizing the conservation aspect of recycling and valuing the social and environmental service of recyclers. Several questions arise: Why do we create products to become waste, in the first place? How can consumption become less wasteful? Can waste recovery be valued as a beneficial environmental service similar to other public services for the good of society?

Why Fight over Waste?

A new conflict that needs to be addressed by governments and the community is the contested ownership over wasted resources. Conflicts can happen between different users of the same 'trap-line,' the geographic routes recyclers generally take to search for material. Official or private agents of recycling programmes sometimes treat informal recyclers or binners as thieves when they collect out of the disposed household waste (Tremblay 2007). Should priority be giving to income generation for the maintenance of many livelihoods or should a few be profiting from the privatization of waste management? There are obvious economic opportunities in waste management, as outlined by Johnson (2004). Davies (2002) speaks about large corporations as being the 'mafia' dominating the international waste management sector. Multinationals like Onyx (Vivendi Environment, France), Vectra and Shanks (UK), Sita (Suez, France), RWE and Rethman (Germany) currently have the largest share of the global waste management market (Davies 2002). Their domain is rapidly expanding throughout the world, and their proposed technical solutions usually are not concerned with redistributing income and generating employment.

Yet unemployment is one of the major problems today. It is also the driving force for many other social problems including crime, drug abuse, violence and social exclusion. Solutions focusing only on technology and economic efficiency are not enough to tackle the complex social questions. Particularly in poor countries, organized and informal recycling provides income to a large number of people. Waste legislation can target social inclusion as it does resource conservation and economic viability. Employment intensity, defined by the International Labor Organization in terms of employment elasticity with respect to output (Khan 2004) is a valid indicator to be considered in decision-making.

The term 'waste management' has been analyzed and re-defined by Eva Pongracz and Veikko Pohjola as 'control of waste related activities with the aim of protecting the environment and human health, and encouraging resource recovery' (2004, 152). To minimize environmental health risks, the hazardousness of waste must be evaluated by environmental government authorities (such as Environmental Protection Agencies) and official development planning agencies and specific actions need to be prescribed by the government as the ultimate regulating body in public health. I suggest a more radical approach, one that not only encourages but mandates resource recovery as the necessary end of the product life-cycle, and that implements inclusive waste management as a contribution to social justice. By inclusive waste management I mean strategies that involve the recyclers organized in associations, co-operatives or other forms of community organizations in selective waste collection. This entails building the capacity of recyclers towards the enhancement of resource recovery, such as door-to-door and other selective collection modes. Informed and trained recyclers have the potential to be environmental stewards. A further step is transforming recyclables into new products, adding value to the materials and expanding the capacity of generating income for the poor. How can we achieve such a deeply rooted ethical change given the current accelerated growth pace and far-reaching market economy? Some of the examples discussed in this book provide insights to practical actions and policies that enhance sustainability.

Cities: The Big Generators of Household Waste – Problems and Opportunities

Urban lifestyles generate significantly more solid household waste than rural livelihoods. The increased food packaging generated by the expansion of industrialised and frozen food, take-away and fast food is just one of many problems in urban centres today. Urban lifestyles further persuade consumption intense and wasteful attitudes. With economic growth and population increase, the problems related to solid waste are most likely going to further increase. Over the past four decades urban growth has been exponential in parts of the developing world. Latin America has already reached the peak with now 75 per cent of the population living in cities. In Africa and Asia a similar trend is underway. Projections for Africa foresee a population change from rural to predominately urban, during just one generation. 36 per cent of the population in Asia lives currently in cities and the population trend is similar with rapidly growing urban populations (UN-Habitat 2007). A historic urban transition is underway, particularly in parts of Asia and Africa with cities experiencing massive population influx from rural areas, resulting in precarious housing and insecure working conditions. The driving motor for this change is the expectation for a better quality of life in the city. Nonetheless, the final destination for many is the urban periphery, where sanitation is inadequate, living conditions are crowded and tenure is insecure. In Brazil exponential urban change has been going on for over 40 years, and today over 80 per cent of the population already lives in cities. Here the living conditions are hazardous, similar to the precarious situation in the poor neighbourhoods in many developing countries. In the city poor and rich consume industrialized and packaged goods. Worldwide solid waste has become a pressing environmental health predicament.

The UN-Habitat (2007) estimated that in 2001, 924 million people, or 31.6 per cent of the world's urban population, lived in precarious housing conditions (*favelas* or *barrios populares*). In 2005, 41.4 per cent of the urban population in developing countries, compared to 6 per cent of the urban population in developed countries, lived in poor neighbourhoods. If no concrete action is taken to halt the situation the total population living in precarious housing in the world is estimated to be over 1 billion people, a number that may increase to 1.5 billion by 2020, according to the United Nation HABITAT organization.

A visible process of urbanizing poverty is currently underway, with growing absolute numbers of poor and malnourished in the cities. Structural adjustment programmes imposed by lending agencies like the World Bank or IMF enlarge existing inequalities, since these programmes tend to reduce the country's social spending. The long *Kondratieff cycles* driven by technological innovations and the shorter *Kuznet boom and bust waves* in the development of the financial and capital markets result in economic restructuring, often enhancing the uneven distribution of wealth. The result is a sharp division between the *haves* and the *have nots*, with extensive poverty often next to wealth (UN-Habitat 2007).

Whereas the percentage of the population considered poor in Asia is declining, it is increasing in Africa and also slightly in Latin America. Poverty is defined primarily as the state of economic deprivation, and the term *social exclusion*, which includes economic, political and spatial exclusion as well as in terms of stigma and discrimination. Social exclusion also denominates the state of being left out from the satisfaction of basic needs, including the access to public services, goods, activities, or resources, which are essential for a life in dignity (Room 1995). Restraining power and disabling people from being full citizens means excluding them. Citizenship is closely tied to human rights. It relates to the wider role of people and their activity in society. 'Citizenship is a key expression of the relationship between the state and the individual and of the individual to society' (Parker 2002, 16). Social exclusion means lack of citizenship. People are stripped of basic citizenship rights, including access to housing, formal education and health care. While poverty and social exclusion are visible and widespread in the South, they are also present, though less obvious, in wealthy societies in the North. There are differences, though, in the extent of the phenomena. In Brazil almost 30 per cent of the population lives under conditions defined as social exclusion – poor housing and living conditions, inadequate nutrition and unsafe working environments. Extensive illegal land occupations, also called squatter settlements or *favelas*, are common, as are under-serviced neighbourhoods, particularly in the urban fringe of large cities. The presence and power of drug traffickers in these neighbourhoods has often eroded local initiatives such as residents associations, and the drug magnates exercise power over the community – the so-called *poder paralelo* (parallel power) (Wheeler 2003). Many of the *catadores* (recyclers) and their families live in *favelas*, often in precarious living and housing conditions.

The growth of informal settlements is usually also accompanied by the growth of the informal economy. In many cities in the developing world, more than 60 per cent of the urban employment is in the informal sector. Informal activities are characterized as coping strategies and as '…active strateg[ies] utilized by marginalized populations to "get by" in highly precarious economic circumstances' (Leonard 2000, 1070). The informal economy is described as unregulated employment or unregistered economic activity (UN 2005). The ILO refers to *decent work deficits*, which are more pronounced in the informal sector, where unsafe and unhealthy working conditions, long working hours with insufficient and unsteady compensation, low skill and productivity levels, and a lack of access to information, markets, finance, training and technology prevail (ILO 2002). Nonetheless, Fernando de Soto recognizes the informal sector as the *engine of growth* due to its economic and social significance (Soto 1989). For countries in Sub-Saharan Africa the situation is extreme, with 78 per cent of the urban employment being informal and this sector generating 42 per cent of the GDP. For this part of the world the UN-Habitat foresees that future growth with small-scale enterprises in the informal sector will provide over 90 per cent of the additional jobs in urban areas over the next decade (UN-Habitat 2007). Informal recycling is always one of the options for individuals and families with no income.

Census data from 1996 confirms that 15 per cent of the working-age population in the city of São Paulo was making its living through informal activities (IBGE 2000). Selective waste collection is one of the widespread informal survival strategies of the *catadores*, the term for informal recyclers in Brazil. The activity involves the recovery of resources, which would otherwise end up in the environment, either deposited irregularly or at landfills. Material is collected out of the garbage, from businesses or households, and is sold to middlemen or directly to industry. Recycling generates income and improves environmental health. Often children and elderly people also participate in this activity. The survival of thousands of people depends exclusively on accessing wasted materials, despite the fact that this work is economically undervalued and the workers face serious health risks and harsh social stigmatization. Severe and chronic occupational health problems are very common among the recycling workers (Gutberlet and Baeder 2007).

Recycling provides also many environmental benefits. Recovering resources from the waste stream helps address global climate change by diminishing greenhouse gas emissions (Pongracz and Pohjola 2004). Although recycling also produces emissions, these are offset by the reduction of fossil fuels that would be required to obtain new raw materials. Recycling reduces the effects of climate change in many ways. Resource recovery diminishes and eventually eliminates the need for landfills (which release methane, a greenhouse gas 20 times more potent than carbon dioxide) and incinerators (which waste resources that could be reused or recycled and incineration generates dangerous emissions and residues) (Forsyth 2005). Decreased demand for virgin paper leaves more trees standing, which increases carbon sequestration in forests. More environmentally friendly however is the voluntary approach that consists in avoiding and minimizing the generation of waste. Product redesign and the redesign of our lifestyles can become the truly significant measures to save energy and resources.

It is not widely recognized that the people collecting recyclables are providing an important environmental service to the public (Ali 2006, Rouse and Ali 2001). Recyclers often have established partnerships with the local community or businesses to collect waste material. In building their capacity as environmental stewards, the collecting and the relationship building can be effective in educating the population towards wasting less and recovering more. With training and diffusion of information the recyclers are strengthened and the activity becomes less hazardous and also less stigmatized. The important environmental contribution that recyclers make must be recognized in the debate over cost recovery and revenue generation from waste management. Although large and wide-ranging waste enterprises may be able to reduce costs with economies of scale, co-operative schemes and community-based enterprises provide the chance for environmental education. In terms of technology investment and maintenance costs, co-operative arrangements and community-based recycling enterprises are cheaper than large-scale operations even considering economies of scale (Ali 2003). The community waste sector in the UK seems to be financially autonomous in delivering integrated waste management and environmental education. However, Luckin and Sharp (2006) point out that usually community waste recycling is lacking the capacity to provide service over a larger area. Market fluctuations and policy changes related to waste handling seem to be the biggest hurdles.

New Social Movement Emerging from the Recycling Scene

Although the level of organization and networking is growing in Latin America, *catadores* continue to be the most excluded, most impoverished and most disempowered segment of society. They are stigmatized for their activity and they suffer under strong societal prejudices. In 1999 a new social movement the Movimento Nacional dos Catadores de Materiais Recicláveis (MNCR) became formalized in Brazil and two years later, gaining momentum on a national level, more than 1,700 recyclers met at their first nationwide congress to discuss their livelihood conditions and policies for change. The MNCR helps to organize and structure the workers into an independent, efficient service sector. The movement promotes regional meetings to strengthen the educational formation of the recyclers and to strengthen the level of organization of the groups and the movement. The establishment of regional committees aims at: 1) Providing conditions for recyclers to exchange their experiences and to plan collaborative actions. 2) Promoting the relation between national commission and the regional groups of recyclers in order to establish so-called 'organic bases' (*Bases Orgânicas*). 3) Fostering the regional networks by creating state co-ordinations of the recyclers (MNCR 2007). The third Latin American 1st World Conference of Informal Recyclers (*Waste Pickers Without Frontiers*) was held in Bogotá Colombia at the beginning of March 2008.

There is hope that organized recyclers can make a difference to environmental health. Many of their members are already part of recycling co-operatives, associations and other forms of organization. Co-operative recycling practices can be part of a strategy to reduce urban poverty. There are numerous examples of collective and independent entrepreneurships of collecting, separating and transforming recyclables everywhere in Brazil. The network is now also collaborating with movements from other countries, particularly within Latin America (Medina 2005).

Over the past years many new recycling co-ops and associations have found new ways of conducting this recycling activity collectively. Experiences from Brazil and other countries demonstrate that individuals become empowered by being part of organized and structured recycling groups, as opposed to working as informal autonomous recyclers without a support network. Lately these groups are consolidating into a larger network to commercialize material collectively. There are also examples of co-ops that have developed strategies to further add value to their material, through artwork or the creation of new products. These experiences provide the ground for a new form of economy, in which profit making is typically not the exclusive or even primary purpose. *Solidarity economy* and *social economy* are defined as '…initiatives and organizational forms – that is, a hybridisation of market, nonmarket (redistribution) and non-monetary (reciprocity) economies showing that the economy is not limited to the market, but includes principles of redistribution and reciprocity' (Moulaert and Ailenei 2005, 2044). Social economy is understood as social innovations, new forms of making business, based on co-operative values and solidarity. Through this form of economy we learn new ways to create and sustain participatory democracy (Moulaert and Nussbaumer 2005). Selective collection as co-operative activity is an innovative example of social economy. The current federal government in Brazil is strengthening the experiences in solidarity economy and promoting social development (Gaiger 2004, Singer 2003). The National Movement of the Recyclers (MNCR) is one of the beneficiaries of these policies.

Participation and Citizenship: The Keys?

It is widely claimed that participatory approaches in planning, action and research can be a solution to the pressing social, economic and environmental predicaments of our cities today. There are nonetheless different forms and levels of participation, varying from consultation to deliberation, which need to be considered. Participatory processes are complex *per se*, because they usually involve different stakeholders, sometimes with opposite views and experiences. In their edited collection, *Participation: The New Tyranny* (2001), authors Cooke and Kothari criticize participatory approaches as dictating procedures, not necessarily contributing to the empowering outcomes claimed in its rhetoric. They point out the malleability of participatory methods, which can allow the facilitator to dominate and manipulate. Furthermore they question the lack of critical analysis at the local scale, because it does not highlight the unequal power relations and romanticises the local, and at the same time ignores broader, structural forms of injustice. Structure and agency in social change were not adequately incorporated into participatory development thinking and practice. However, participatory processes involve 'a rethinking of development, political community, and democracy and require a…"critical modernist" approach to development' (Hickey and Mohan 2005, 1).

Critique of participatory approaches is justified. In particular, radical 'insider' voices early on critiqued the adoption of participation by powerful development agencies (Rahman 1995). 'The current recasting of participation within the frameworks of "empowerment", "democratic governance", "rights-based approaches" and "social accountability" reveals, on closer inspection, little more attention to the underlying causes and power effects of poverty and inequity than in previous incarnations' (Cornwall 2006, 78). The author analyses the various approaches to participation and their impacts over the past. She finds '…not only a familiar set of arguments and practices, but a fairly narrow set of governance problematics, which development actors have addressed with repeated calls for greater participation' (Cornwall 2006, 78). The historical analysis of how participation has been used within development reveals it as an 'artefact of plural cultures, articulated at the interface between the organizational cultures of development agencies and the imperial cultures that shaped the very notion and practice of "development"' (Cornwall 2006, 79).

Participatory and citizenship discourses touch on underlying power relations. Within these discourses the focus is on existing participatory forms of citizenship (for example, social movements), which aim towards social transformation, expressed through proposing critical alternatives rather than the abandonment of participatory political action (Hickey and Mohan 2004). Participation has been embraced as a way to build transparency, accountability and trust between people and institutions. It has almost become a mainstream approach by researchers, funding agencies, and governments to tackle poverty and social injustice. By strengthening citizen rights and providing spaces where citizens have a voice, participatory development also becomes a basis for personal learning and change, which can help produce a wider societal transformation. The participatory approach of institutional actors and popular agency also creates new forms of governance. Here participation is seen as citizenship. 'The aim is to transform the political process in ways that progressively alter the processes of inclusion and exclusion that operate within particular political communities, and which govern the opportunities for individuals and groups to claim their rights to participation and resources' (Hickey and Mohan 2005, 3). Prime examples include social movements that claim land and shelter, such as the landless or homeless movements in Brazil, and those that claim both cultural and political rights for members, as well as participatory budgeting, already a widespread reality in Brazil. The challenge to ensure a transition from clientelism to citizenship remains critical.

Lister connects participation with citizenship analysis, placing it into a wider socio-political context (1997). Participation enables people to extend their status as members of particular political communities, thereby increasing their control over socioeconomic resources. Other discourses highlight the pro-active notion of participation, where citizenship can be claimed from below through the efforts of the marginalised in organised struggles, rather than waiting for it to be granted 'from above' (Lister 1997). In participatory processes there is a chance for people to hang on to decisions and share responsibilities exactly because they have been involved in the process, they have participated in the debate rather than being obligated by law. 'Such a position retains a belief in the central tenets of modernism – democracy, emancipation, development and progress – while allowing for the multiple ways in which these processes can be engaged, and contended through deliberative forms of democracy' (Hickey and Mohan 2005, 4).

However, Hickey and Mohan affirm: '…Much participatory development focuses on recognition of marginalised groups without a concomitant commitment to redistribution. This links, in some ways, to the means and ends of participation. Focusing on participation as an end does not guarantee pro-poor redistribution. Overcoming marginalisation requires institutionalising recognition and redistribution together, within a broader project of social justice' (2005, 4).

Top-down, expert led processes are less likely to construct sustainable communities. Experts alone are unable to deal with the complexity of sustainability. Citizens need to be involved, because in the end they will be affected by on the ground results. Mechanisms such as alternative governance arrangements need to be provided for effective citizens participation. Latin America has been in the media for some time regarding new forms of governance. Brazil in particular has become a leader in terms of innovative and creative social development. Participatory public policies, such as *orçamento participativo* (participatory budgeting) and democratic urban planning can make a difference to the citizen's quality of life. Participatory budgeting was introduced in Porto Alegre in the early 1990s and is now a widespread common practice in many cities in Brazil (Abers 1998). These forms of political structures can ensure a wider active political and economic participation of the marginalized. With a growing population of urban poor almost everywhere, innovative solutions are urgently needed. Many examples indicate that ignoring social-economic problems instead of tackling them only postpones and increases the risk of conflicts and social upheaval.

Participatory waste management contributes to social cohesion and builds stronger communities. Networks are platforms to communicate and exchange experiences, which can be used as discussion forums to help improve occupational safety and increase income generation The scope of the unresolved waste dilemma underlines the urgency of new arrangements in waste management. New policies are required that address the waste of energy and natural resources and the increased generation of garbage in our disposable society. Strategies that on one hand promote integrated selective collection, separation and recycling and on the other hand generate income for the urban poor will be most suitable. Participatory resource management or co-management are well known practices in the context of natural resources management, but no experiences have been documented on the use of these approaches in waste management. This book suggests the transfer and adoption of these co-arrangements to this urban sphere.

Theoretical Structure

The theoretical framework underlying this work uses elements that stem from Social Theory and from Political Ecology (Figure 1.1). Community development, social cohesion, social inclusion, participation and empowerment are key concepts in social theory that inform different perspectives of this debate (Armitage 2005). Specific questions look to political ecology for answers to the social and environmental injustices and the emerging distributive and public policy issues (Bullard, 1990). The main themes that permeate this book are related to solid waste generation, urban consumption and participatory resource recovery as social and environmental solution for the current waste dilemma. I will discuss the question of inclusive recycling basically from three interwoven approaches: 1) social and solidarity economy, 2) resource management and 3) governance. These concepts will re-emerge throughout the discussion.

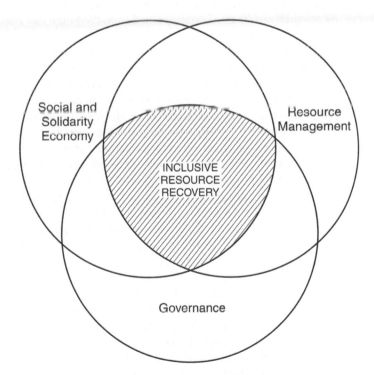

Figure 1.1   Theoretical framework for inclusive waste management

Source: Jutta Gutberlet.

Solid waste is as much a resource as other natural commodities for which there is supply, demand and a price. As an open-access resource, conflicts can derive over its recovery, particularly in situations of scarcity. Management guaranteeing fair access to the resources and fair distribution of the benefits are crucial. In bringing together social economy with resource management, the book tries to evoke new avenues of thinking, proposing an integrated approach that addresses social and environmental challenges related to the generation and management of solid waste. The concepts of participatory or co-management (Berkes and Folke 1998, Ackerman, 2004) are central in the debate about inclusive public policies for solid waste management. Social economy as being community-driven rather than profit-driven, highlights the social gains from such a concept (Mulaert and Nussbaumer 2005). Participatory governance structures are crucial in order to promote long-term and widespread change towards more responsible consumption patterns and more adequate waste management schemes (Forsyth 2005, Zwart 2003). The book will present innovative policy frameworks from Brazil that facilitate inclusive waste management and hence contribute to poverty reduction, environmental quality and the building of stronger communities.

Through the overlap of these theoretical lenses we are able to understand the complexity of our social, economic and environmental problems linked to the production of waste and we can envision resolutions to some of our problems. I hope that some of the research results and lessons learned from the South can also inspire solutions in the North and vice versa.

The book focuses on the situation in Latin America, particularly on several case studies from the metropolitan region of São Paulo in Brazil (Figure 1.2). In Chapter 2 the general predicaments of urban growth, consumption and increase in solid waste are discussed as one of the most pressing current problems in cities in the South and in the North. Solid waste is a threat to human security. However, it can also be treated as a livelihood opportunity for the urban poor. New global perspectives in what to do about our growing mountains of solid waste will be introduced in this section.

Chapter 3 examines some of the social and economic conditions of marginalized urban dwellers. A case study illustrates the dimension of social exclusion in a squatter settlement in the southern fringe of São Paulo. At the time of the research there was no legal and safe access to water and electricity; sewage ran in open ditches and waste was irregularly dumped, as it is still widespread in other neighbourhoods in São Paulo, throughout Brazil and in the developing world in general. Population density is extremely high, whereas green spaces and leisure areas are mostly inexistent. The case of the community *Pedra sobre Pedra* shows the pressing need to tackle waste in marginal settlements as a prime way to enhance liveability.

In Chapter 4 I discuss the production-consumption framework and resulting environmental health impacts from the generation of waste. The worldwide increase in resource-intense consumption and in unsustainable lifestyles over the past two decades, has significantly added to the problems of garbage generation. Illegally deposited waste is a particular problem in the urban periphery of large cities in the South. In some communities, local neighbourhood associations have initiated alternative solutions for their garbage problem. After highlighting these predicaments, communal recycling initiatives are discussed in terms of their assets and barriers for success.

Figure 1.2   Localization map metropolitan region of São Paulo

Sources: CPLA/SEMA 1997 and IBGE 1991.

Cartography: Ole J. Heggen.

Occupational health risks that come along with selective waste collection and separation are discussed in Chapter 5. During their work informal and organized recyclers are exposed to severe physical, chemical and biological hazards. The working conditions often imply high risks of accidents, besides reinforcing socio-economic exclusion and stigmatization, which also affects the health of the recyclers. A case study on the situation in Santo André, a city next to São Paulo, reveals numerous health complaints and stories about the daily hardships of collecting recyclables in the streets. The results underline the fact that policy makers at all government levels urgently need to address these issues, which affect so many people throughout Brazil and abroad.

Chapter 6 highlights the increasing role of selective waste collection and recycling for poverty reduction and resource recovery, improving environmental quality. Overall, urban poverty has increased during the past decades attracting more and more people to the informal selective waste collection and separation. The chapter discusses new ways of organizing this activity, such as through recycling co-operatives and associations. Approaches based on integrated solid waste co-management with recycling co-operatives or associations are increasingly becoming instrumental in improving the livelihoods of recyclers. The total extent of the valuable contributions from recycling co-operatives and from the mostly informal recycling sector is not yet fully recognized. There are major political and economic barriers that still prevent inclusive and participatory public policies in waste management from happening.

The municipalities Diadema and Ribeirão Pires in the metropolitan region of São Paulo showcase some of the successes and hurdles of organized recycling. The need for a paradigm shift toward recognizing the social and environmental benefits of this activity, involving the Government and the public, is urgently needed. The results presented here are contributions towards wider pro-poor policy making.

The concluding chapter, Chapter 7 strengthens the case for participatory approaches as solutions to current urban social, economic and environmental predicaments. With a growing population of urban poor almost everywhere, innovative solutions are urgently needed. Examples indicate that ignoring social-economic problems instead of tackling them only postpones and increases the risk for conflicts and social upheaval. Participatory waste management can contribute to social cohesion and hence build stronger communities. The livelihoods of the recyclers can be significantly improved with adequate integrated policies.

Closing the loop with recycling is not a final solution. It can, however, be an approach to preserving virgin resources, to wasting less and to generating environmental and social awareness. Usually, the poorest and most excluded individuals of society make a living through gathering the valuables out of our garbage bins; yet they are passed over and ignored in waste management. Our waste and also the practice of over-packaging and disposability – where everything is easily disposed after one time use – is being transferred to other regions in the world, causing environmental contamination and resource depletion, even in remote rural areas. Waste incineration is globally widespread technology. This solution to waste needs to be immediately evaluated, last but not least because of the social-economic impact it causes by displacing workers from the formal and informal waste management sector. We must also remember that waste incineration causes the emission of PCBs, dioxins, furans, and other toxic materials, adding to global climate change (Morris 1996). Consumption is at the heart of the problem. We need to reconsider what we consume, why we consume, and how we can reduce our consumption levels. Our actions directly affect the environment and society, sometimes immediately, sometimes cumulative and later. It is our choice whether to pay the real price now or whether to pay for the omitted externalities once they have grown into larger sustainability issues. Waste is a cross-cutting theme that tackles social and environmental justice from the local to the global scale and that permeates into different time scales.

Finally, I would like to bring to our attention the use of the word development, which is widely identified with the processes of economic growth and industrialization. From that perspective, *developing countries* means those countries that are either in the process of industrializing or are still primarily involved in rural activities; usually these are countries with large income inequalities. The term is criticized for implying that industrializing is the only way forward. I want to stress that economic development is frequently not the most beneficial model, and objections to the terms *development* and *developing country* are too important to be ignored (London 2003). The *Global South* and *majority countries* are new terms to describe the world's poor countries. *Global South* implies neither progress nor retreat; it also captures the power differentials and unequal exchange that have historically and undeniably characterized relations between *northern* (that is, industrialized) countries and the rest of the world.

References

Abers, R. (1998), 'Learning democratic practice: distributing government resources through popular participation in Porto Alegre, Brazil', in Douglas, M. and Friedmann, J. (eds), *Cities for Citizens* (Chichester: Wiley & Sons).

Ackerman, F. and Mirza, S. (2001), 'Waste in the Inner City: Asset or Assault?', *Local Environment* 6(2), 113–20.

Ali, M. (2003), 'Community-based enterprises: constraints to scaling up and sustainability', *Solid Waste Collection that Benefits the Urban Poor*, Vol. 2. (Dar Es Salaam).

——— (2006), 'Urban waste management as if people matter', *Habitat International* 30(4), 729–30.

Cooke, W. and Kothari, U. (eds), (2001), *Participation: The New Tyranny?* (London: Zed books).

Cornwall, A. (2006), 'Historical perspectives on participation in development', *Commonwealth and Comparative Politics* 44(1), 62–83.

Davies, S. (2002), 'Waste management multinationals', *Public Services International Research Unit* (PSIRU), (School of Social Sciences: Cardiff University).

Denison, R.A. (1996), 'Environmental Life-Cycle Comparisons of Recycling, Landfilling and Incineration', *Annual Review of Energy and Environment* 21, 191–237.

Drackner, M. (2005), 'What is Waste? To Whom? – An Anthropological Perspective on Garbage', *Waste Management and Research* 23, 175–81.

Etzioni, A. (1988), *The Moral Dimension: Toward a New Economics* (London: Free Press).

Forsyth, T. (2005), 'Building deliberative public-private partnerships for waste management in Asia', *Geoforum* 36, 429–39.

Friedman, M. (1996), 'A positive approach to organized consumer action: The "buycott" as an alternative to the boycott', *Journal of Consumer Research* 19, 439–451.

Gaiger, L. (Org.) (2004), *Sentidos e Experiências da Economia Solidária no Brasil* (Porto Alegre: Editora da UFRGS).

Gutberlet, J. and Baeder, A. (2007), 'Informal recycling and occupational health in Santo André, Brazil', *International Journal of Environmental Health Research* 18(1), 1–15.

Hartwick, E. (2000), 'Towards a Geographical Politics of Consumption', *Environment and Planning* 32, 1177–92.

Hempel, L.C. (1999), 'Conceptual and analytical challenges in building sustainable communities', in Mazmanian, D.A. and Kraft, M.E. (eds), *Towards Sustainable Communities* (Cambridge, MA: MIT Press).

Hickey, S. and Mohan, G. (eds), (2004), *From Tyranny to Transformation? Exploring New Approaches to Participation* (London: Zed Books).

—— (2005), *From Tyranny to Transformation?*, Participatory Development Forum Conference: 'Participating to Create a Different World: Shaping Our Own Future' (University of Ottawa: Canada).

IBGE (Instituto Brasileiro De Geografia e Estatística) (2000), *Contagem da População 1996*, 2000 (Census data 1996) São Paulo: Fundação Instituto Brasileiro de Geografia e Estatística, [website] http://www.ibge.gov.br, accessed 1 October 2007.

ILO (International Labour Organizaion) (2002), *Women and Men in the Informal Economy: A Statistical Picture* (Geneva: ILO office).

Johnson, C. (2004), 'Uncommon ground: The "poverty of history" in common property discourse', *Development and Change* 35(3), 407–33.

Khan, A.R. (2004), 'Growth, Inequality and Poverty: A Comparative Study of China's Experience in the Periods Before and After the Asian Crisis', *Issues in Employment and Poverty*: ILO Discussion Paper 15.

Leonard, M. (2000), 'Coping strategies in developed and developing societies: The workings of the informal economy', *Journal of International Development* 12, 1069–85.

Lister, R. (1997), *Citizenship: Feminist Perspectives* (Hampshire: Macmillan).

London, D.J. (2003), 'The economic context: grounding discussions of economic change and labor in developing countries', in Wokutch, R.E., Hartman, L.P. and Arnold, D.G. *Rising Above Sweatshops: Innovative Approaches to Global Labor Challenges* (Westport, Connecticut: Praeger), 49–76.

Luckin, D. and Sharp, L. (2006), 'The community waste sector and waste services in the UK: Current state and future prospects', *Resources, Conservation and Recycling* 47, 277–94.

Maser, C. (1997), *Sustainable Community Development: Principles and Concepts* (Florida: St. Lucie Press).

Medina, M. (2005), 'Serving the unserved: informal refuse collection in Mexico', *Waste Management and Research* 23, 390–7.

MNCR (Movimento Nacional dos Catadores), [website] http://www.movimentodoscatadores.org.br, accessed 6 July 2007.

Morris, G. (1996), 'Recycling versus Incineration: An Energy Conservation Analysis', *Journal of Hazardous Materials* 47, 277–93.

Moulaert, F. and Ailenei, O. (2005), 'Social Economy, Third Sector and Solidarity Relations: A Conceptual Synthesis from History to Present', *Urban Studies* 42(11), 2037–53.

Moulaert, F. and Nussbaumer, J. (2005), 'Defining the Social Economy and its Governance at the Neighbourhood Level: A Methodological Reflection', *Urban Studies* 42(11), 2071–88.

Parker, G. (2002), *Citizenship, Contingency and the Countryside* (London: Routledge).

Pongracz, E. and Pohjola, V. (2004), 'Re-defining waste, the concept of ownership and the role of waste management', *Resources, Conservation and Recycling* 40, 141–53.

Rahman, M.A. (1995), *People's Self-development: Perspectives on Participatory Action Research* (London: Zed Books).

Robins, N. and Kumar, R. (1999), 'Producing, providing, trading: manufacturing industry and sustainable cities', *Environment and Urbanization* 11(2), pp. 75–93.

Room, G. (ed.) (1995), *Beyond the Threshold: The Measurement and Analysis of Social Exclusion* (Bristol: The Policy Press).

Rouse, J. and Ali, M. (2001), *Waste Pickers in Dhaka: Using the Sustainable Livelihoods Approach – Key Findings and Field Notes*, Water, Engineering and Development Centre, Loughborough University.

Schumacher, E.F. (1999), *Small is Beautiful: Economics as if People Mattered: 25 Years Later*, 3rd edn (Point Roberts, WA: Hartley & Marks Publishers).

Shaw, D. and Newholm, T. (2002), 'Voluntary simplicity and the ethics of consumption', *Psychology Marketing* 19(2), 167–85.

Sheehan, B. and Spiegelman, H. (2005), 'Extended producer responsibility policies in the United States and Canada', in *Governance of Integrated Product Policy* (Sheffield: Greenleaf Publishing).

Singer, P. (2003), 'As grandes questões do trabalho no Brasil e a economia solidária', *Proposta* 30:97, 12–16.

Soto, de H. (1989), *The Other Path: The Invisible Revolution in the Third World* (New York: Harper & Row).

Tremblay, C. (2007), *Binners in Vancouver: A Socio-economic Study on Binners and their Traplines in Downtown Eastside*, Master of Arts Thesis, Department of Geography, University of Victoria, Canada.

UN (United Nations) (2005), *The Inequality Predicament: Report on the World Social Situation 2005* (New York: United Nations).

UN-HABITAT (United Nations Human Settlement Programme) (2007), *State of the Worlds Cities 2006/7: The Millennium Goals and Urban Sustainability* (Nairobi: UN-Habitat).

Weber, E.P. (2003), *Bringing Society Back In: Grassroots Ecosystem Management, Accountability and Sustainable Communities* (London: MIT Press).

Wheeler, J.S. (2003), 'New forms of citizenship: democracy, family, and community in Rio de Janeiro, Brazil', *Gender and Development* 11(3), 36–44.

Zwart, I. (2003), 'A greener alternative? deliberative democracy meets local government', *Environmental Politics* 12(2), 23–48.

Chapter 2

Cities, Consumption and Disposable Society

Introduction: Urban Growth and the Challenges in Waste Management

Contemporary cities as products of the post-industrial consumer society can be defined as spaces of consumption, and it is consumption that heavily shapes urban development (Jayne 2006). Rapid urban population increase brings about urban sprawl, leading to physical changes in the landscape. Consumption is evident almost everywhere in this new landscape: it involves the production, advertisement, transportation and commerce of products, and as a consequence the generation of waste throughout the product life-cycle. Cities shape consumption and consumption shapes the city. Spaces are created primarily for consumption, targeting real needs and created or fashioned wants. Urban spaces are complex and it would be superficial to generalize about them, particularly when comparing diverse cultural settings. There are obvious differences between the modern and post-modern cities in the North and in the South; and yet some of the problems are similar in both hemispheres.

We are now living in a predominantly urban world. According to the United Nations, by 2030 approximately 5 billion people (60 per cent) will be living in cities. At the same time it is predicted that the rural population will decline from 3.3 to 3.2 billion people (UNFPA 2004). The urban drive brings significant changes in lifestyle and the adoption of new consumption patterns. In Africa and Asia, urban agglomerations are growing twice as fast as the overall population growth rate. Each day, approximately 160,000 people migrate from rural to urban areas on these continents. The estimated urban growth rate for high-income, so-called *developed countries* is 0.5 per cent, compared to 2.7 per cent in low-income or *developing countries* and 4.5 per cent in the poorest nations, the so-called *least developed* countries (UNFPA 2001). Urban transformation into *city-regions* and growing *metropolitanization* are current trends, particularly in the South, with an increase of *metacities*, massive conurbations of more than 20 million people, a concept that exceeds the previously largest *megacities* (UN-HABITAT 2006, 6). At the same time these cities are experiencing unprecedented poverty and environmental health impacts because of water, air and soil contamination. Through the increasing accumulation of solid waste, the social and environmental dimensions of these issues become visible.

In this chapter I analyze the current challenge of increased solid household waste in large cities. The discussion focuses on rapid global environmental change and its influences on human security. Since the very early population increase humankind has impacted on the environment; however it is the accelerated pace and the global extension of it that is novel and so problematic. 'As a result of human actions, the structure and functioning of the world's ecosystems changed more rapidly in the second half of the twentieth century than at any other time in human history' (WHO 2005, 6). Urban growth is synonymous with increase in waste. What are some of the key environmental and social issues associated with waste and what are some of the alternatives in its management from a social-environmental perspective? Recycling is often seen as one of the solutions to increased solid waste, besides other more conventional methods like incineration and sanitary landfills. The obvious limitations of these practices are mostly overlooked. These issues are central in the debate on sustainable cities and healthy communities. I will introduce São Paulo, as an example of a megacity with growing solid waste accumulation, a typical current scenario in large metropolitan agglomerations.

The concept of human security has been introduced by UNDP as a universal, preventive approach focusing on *freedom from fear* and *freedom from want* (UNDP 1994). More specifically the International Human Dimensions Program defines human security as 'a state that is achieved when and where individuals and communities have the options necessary to end, mitigate or adapt to threats to their human, environmental and social rights; have the capacity and freedom to exercise these options; and actively participate in pursuing these options' (GECHS 2007). It is pertinent to underline the link this concept makes between poverty and insecurity as well as its multidimensional and people-centred nature. This approach is used in research to identify, understand and propose resolutions to human development issues. In particular, it concentrates on how environmental change impacts and threatens individuals and communities. In what kind of preventive actions can people engage and actively participate to pursue human security? Environmental change in the context of urban growth can impact the security of individuals and communities in many different ways. Hence, the objective of human security is to safeguard the vital core of all human lives from critical pervasive threats, without impeding long-term human fulfillment (Alkire 2003).

Rapid urban expansion, energy-intensive production, and resource-intensive lifestyles are triggering global environmental change and posing threats to human security. Along with this growth come environmental degradation, water, soil and air pollution and enormous amounts of waste. The transition from rural to urban livelihoods, particularly in the Western world, often means the adoption of consumption-oriented individualistic lifestyles. The city creates new needs and wants, demanding the extraction of raw materials and energy, as well as generating leftovers from production and consumption. The search for adequate solutions to increased solid waste is among the top challenges in our cities. Though municipalities are facing serious bottlenecks in their waste management systems, in most cases their governments still do not address reduction of waste generation as priority.

One facet to this problem, yet to be properly tackled, is packaging waste. Packaging is classified as all materials for containment, handling, and delivery of products, goods, and raw materials and it includes industrial packaging as well as primary, secondary, and tertiary packaging. Over-packaging is defined as 'unnecessary expenditure on packaging which goes beyond that necessary for protection of the product for transport or storage or beyond consumer requirements' (Pöll and Schneider 1993, 1). Policies are urgently needed to regulate the reduction of unnecessary wrapping,

There are measures to first of all avoid and as a second option to minimise and redirect waste into biodegradability, for reuse, or recycling. Issues that rise here are who is responsible for packaging waste? And who should pay for its disposal?

Despite the social, cultural, environmental, and economic differences between cities in the North and in the South, there are some parallels between all of them when it comes to problems related to the current generation and disposal of waste. Cities worldwide produce much more household and industrial waste than they can manage. Due to the lack of adequate local solutions, the solid residues are usually transferred to different municipalities or even countries (*transboundary shipments*) displacing health risks into other geographical regions, besides adding to the ecological footprint by burning additional fossil fuels with transportation.

In order to reduce the amount of solid waste entering landfills, the recent trend in waste management in the North is to turn to legislation and regulations that formalize municipal recycling programmes (Mee et al. 2004). In the South, most of the recycling happens as part of the informal economy. The large volume of recyclable solid waste combined with high unemployment has resulted in a growing informal recycling economy that is dependent on this resource. Municipalities are caught trying to balance their own political agendas with the demands for sustainable development and improved quality of life by involving private sector initiatives, often multi-national corporations (Baud et al. 2001). Yet the transfer of official waste collection and recycling programmes to the private sector displaces the informal recyclers, compromising their livelihoods (Ahmed and Ali 2004; Berthier 2003; Köberlein 2003).

Many cities, particularly in developing countries still fail to collect part or all household waste. The situation is worst in the poorest regions, particularly in Africa and Asia, where some cities have no official garbage collection. The city of Calcutta, for example, collects about 82 per cent of its waste, whereas other municipalities within the Metropolitan area of Calcutta only collect between 20 and 50 per cent of the waste they generate (Hasan and Khan 1999, 104). These circumstances pose significant health threats to the local population. There are also many examples of municipalities that have initiated progressive recycling programmes. In Chapter 6, I will further discuss public private partnerships in integrated waste management, the bottlenecks and possible benefits.

Consumption and Waste

Statistics Canada (2000 website) defines waste as '...all material unwanted by the generator' and according to the European Union (2006, 5) it is '...any substance or object...which the holder discards or is required to discard'. Furthermore, OECD/ Eurostat (2006) refers to waste as '...materials that are not prime products (i.e. products produced for the market) for which the generator has no further use for own purpose of production, transformation or consumption, and which he [she] discards, or intends, or is required to discard' (website). In the German waste management handbook by Bilitewski et al. (1994, 21) waste is defined as 'portable objects that have been abandoned by the owner'. Basically, 'waste exists where it is not wanted' (Pongracz and Pohjola 2004, 143).

These definitions are rather subjective. Waste is seen as a *state of mind* rather than a definable entity. No explicit economic value seems to be attached to waste once it has been declared waste, and potential social benefits from resource recovery are not recognized in these official definitions. However, the meaning of waste is complex as there are many different types and uses for it, with both positive and negative connotations. Perceptions of waste and the categorization of waste are shifting. Increasingly, recycling programmes generate profits and cities become protective over the recyclable portion of their waste. Consequently, legislation in many countries is changing towards recognizing recyclable waste as a resource with an economic value. This change is also reflected in the statistics on waste for example in Italy, where the recyclable portion of the household waste is no longer included in the data on total household waste. At first glance, the statistics seem to report that less waste had been generated, however, the total amount of waste has not really diminished; part of it has been diverted into recycling.

Standard waste management is defined as the '…control of waste-related activities with the aim of protecting the environment and human health, and resources conservation' (Price and Joseph 2000, 97). Conventional waste treatment focuses primarily on the handling of waste. 'Very little consideration or effort has yet gone into true waste minimization or reduction in the demand that leads to waste generation in the first instance' (Price and Joseph 2000, 97).

Burgess et al. (2003) show the historical evolution of consumption from mercantile capitalism, where consumption beyond the satisfaction of basic needs was mainly for the upper class, driven by their demands for luxuries. After the *English Enlightenment* the ideal that happiness was achievable through consumption, and consumption was available to anyone willing to work hard, was the pathway promoted to reach satisfaction. Mass consumption started in the twentieth century. The *home* emerged as an important locus for consumption activities and numerous items, promoted as essential, were developed to increase the liveability at home. In North America mass consumption rapidly increased during the 1920s and the 1930s with the expansion of industrial mass production. Large corporations, such as Ford and General Electric propagated the drive and the need for households to continuously renew and upgrade their home appliances. Consequently, women as home-makers were particularly targeted in mass advertisements. New consumption patterns brought about changes in the socio-cultural activities, for example, with the emergence of fast food and takeaway, encouraging packaging and disposability (Burgess 2003, 266). Until today no significant step has been made to regulate the packaging industry to avoid and diminish over-wrapping.

Worldwide Differences in Consumption Levels

Under the global expansion of instant mass media and widely spreading capitalism, along with increased mobility and tourism, the disposable character of mass consumption has reached a global scale. Western consumption patterns are transferred to remote locations, impacting on local consumption habits, traditional lifestyles, and local cultures. Widespread consumption and careless disposal of packaging spreads the traces of consumption into the public realm and pristine environments are often threatened by accumulating waste. The following figure (Figure 2.1) highlights two data sets. The first shows that worldwide private and public consumption has grown from US$4.8 trillion in 1960 to US$21.7 trillion in 1995 (UNFPA 2001). The same information was not available for more recent years. The second data set shows only the numbers for private consumption of goods and services at the household level. Public consumption worldwide is not included in this data and therefore it only appears that consumption had decreased in recent years (Worldwatch Institute 2007).

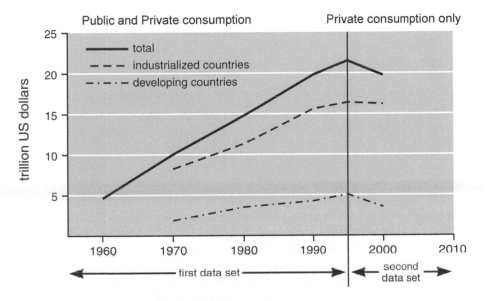

Figure 2.1   Growth in consumption expenditures in developing and industrialized countries 1960 to 2000 (in trillions US dollars)

Sources: UNFPA, The State of World Population 2001 (first data set) and Worldwatch Institute, *State of the World 2004: Consumption by the Numbers* (Second data set).

Cartography: Ole J. Heggen and Jutta Gutberlet.

The disparity of consumption levels between rich and poor countries is large. According to the Worldwatch Institute (2007) 81.6 per cent of the world's wealth is spend in developed countries and 18.4 per cent in developing countries. The United States and Canada are the leading countries in consumption expenditures with 31.5 per cent of the global total, despite only housing 5.2 per cent of the world's population. 33.3 per cent of the world population live in South Asia and Sub-Saharan Africa, yet they consume only 3.2 per cent of the total consumption expenditure. In poor countries there is an equally significant internal disparity between the elite and low- or no-income groups. Access to energy resources and specific consumer goods are used as indicator to show this gap between the wealthiest and the poorest in the world (Table 2.1). Although per capita GDP is only a partial indicator for economic development, it is notable that in 70 countries, the per capita GDP is lower than it was 25 years ago (UNFPA 2001).

Table 2.1   International comparison of consumption levels

	Consumption (in %) by people in industrialized countries	Consumption (in %) by the poorest 20% of people in developing countries
Total energy resources	58	<4
Meat and fish	45	5
Paper	84	1.1
Vehicles	87	<1
Telephone lines	74	1.5

Source: UNFPA, *The State of World Population 2001*, http://www.unfpa.org/swp/2001/english/ch03.html.

Cartography: Ole J. Heggen and Jutta Gutberlet.

Western lifestyles are portrayed by the mass media as models for progress and development, and are spreading quickly into poor nations. One indicator for this change is increased generation of household waste. High consumption levels still generally correlate with intense resource use and waste generation. Waste generation is highest in rich countries, particularly in North America, Australia, New Zealand and Japan with up to two kg/day/capita on average. In most countries in Europe the generation is significantly lower, averaging one to 1.5 kg/day/person. In cities with efficient recycling programmes the amount of waste is visibly less. In Stuttgart, Germany, the average person generates 1.1 kg per day, of which 0.4 kg is recycled. In urban Latin America a person generates on average between 0.5 to one kg of solid waste per day, similar to Hong Kong (1.01 kg/person/day) or Guangzhou in China (1.09 kg/person/day) (Chung and Poo 1998, 207). In low-income countries the values fluctuate between 0.3 and 0.6 kg/day/person; these averages camouflage the widely existing differences in income and consumption levels within the population.

There are also regional and urban/rural differences in waste generation. Statistics usually do not reveal the differences in consumption between rich and poor, nor does the data inform about the level of waste collection. Irregular squatter settlements are often not serviced by formal waste collection. The average production of household waste in Brazil is approximately 0.5 kg per capita per day (UNEP 1999). In large cities like São Paulo, however, the daily per capita municipal waste production is on average 0.85 kg (UNEP 1999). A detailed study on the generation of household waste for Brazil differentiates a generation rate of 0.9 and a collection rate of 0.6 kg on average per person per day (Fehr et al. 2000). The in-depth data also underlines the fact that over the past ten to fifteen years, per capita solid waste generation has increased in almost every city not only in Brazil but around the world (Fehr et al. 2000).

A waste composition analysis conducted by the Japanese International Cooperation Agency (JICA) shows the following proportional waste composition for the city of

São Paulo: 58 per cent organic material, 11 per cent paper and cardboard, 18 per cent plastics, two per cent metal, two per cent glass, and a small proportion of other materials such as batteries, wood, textiles, rubber and undefined materials (JICA 2004, 17). The proportion of plastic in household waste has more than doubled over the past five years. In other countries, such as India for example, recyclable material in household waste has grown significantly from 9.6 per cent in the early 1970s to 17.2 per cent in 1995 (Gupta et al. 1998, 141).

With economic globalisation, consumption patterns in the South are becoming increasingly similar to those in the North. One-way packaging is destined to only do the job of packaging a product once and is then discarded, it may be recyclable or not. One-way, non-biodegradable packaging of food and beverages are on the rise everywhere and are often the only available consumer choices, or the cheapest. However, the disposal of packaging adds to the city's waste management costs, and of course to the depletion of non-renewable resources (oil, gas and minerals) and renewable resources (renewable energy, forests and other plant resources). Environmentally friendly packaging can reduce costs and impacts.

Wasting Food as Part of the Household Waste

Besides recyclables there are other wasted and underutilized treasures contained in household garbage. In some countries more than half of household waste is organic material, which could be composted and reintroduced in agriculture and animal feeding (Mollison 1990; Warwick 2001). In Canada approximately 30 to 40 per cent of all municipal solid waste is composed of organic material (Forkes 2007). The quantity of organic matter indicated by moisture content and biodegradability varies by country. In high-income countries, the moisture content is usually lower (from 20 to 40 per cent), compared to low-income countries (between 40 and 80 per cent) (Cotton et al. 1999, 3). In their study of middle-income condominiums in Uberlândia, Central Brazil, Fehr, de Castro and Calçado (2000) found a very high percentage of organic material in household waste. 72 per cent of the waste was considered biodegradable, of which up to 28 per cent were considered food waste (Fehr et al. 2000, 254). Similar situations regarding food waste are common elsewhere. The most recent waste composition study in the Capital Regional District of Victoria, BC in Canada indicates that over 30 per cent of the solid waste deposited in their local landfill is organic material, of which a significant amount is made up of food waste (CRD 2007).

Food is wasted in rich and in poor countries, in occidental and in oriental cultures and in different socio-economic contexts. The act of wasting happens at different levels and in different scales. At the commercial level, including supermarkets, restaurants and fast food outlets, the food waste per establishment is larger. Occasionally businesses have arranged partnerships with farms that recycle organic waste. On the household level the waste of food is more problematic and depends on the awareness of household members and the ability to compost.

The amount of wasted food depends more on cultural values and attitudes than on income levels. In our disposable society, throwing away is considered a normal practice as an end solution to products and materials that are not wanted or needed any more. Over-consumption – buying more than is necessary to satisfy needs – is common. The consequences in terms of food waste are significant. Food is often not consumed before the expiry date and ends up in the trash bin. In mass-consumption oriented societies this is a widespread and unquestioned practice. In Seattle, for example, it is estimated that approximately 40 per cent of the garbage is food waste. Food waste has become more obvious in North America through the documentation of food scavenging. It highlights an embarrassing unbalance in terms of distribution and access but also in terms of values and attitudes towards resource conservation.

In Vancouver, mapping food waste locations for scavengers has recently drawn public attention towards the issue of wasting and recovering food resources. In July 2006, the Britannia Community Services Centre, a Vancouver-based support group for homeless people, issued a pamphlet showing locations in the city where food is wasted. The polemic document draws attention to two facts: food is wasted and part of the population relies on food that is thrown away. Interestingly the pamphlet reveals hints such as '...Look behind markets, stores and food banks to see what they throw away at the end of the day' (Britannia Community Services Centre 2006). Nevertheless, dumpster food salvaging is unacceptable because first of all it reinforces *throw-away* instead of *give-away* practices; secondly it can cause health problems for the people who consume these products; and thirdly it continues to reinforce undesirable attitudes such as wasting.

Better solutions need to be implemented to prevent food waste, such as transferring resources to food banks and setting up partnerships to provide free meals. In the city of Diadema, Brazil, local food producers and sellers donate products and leftovers to the local food bank, which distributes food baskets to the most needy population. They also contribute to a government-sponsored restaurant, where the no-income or low-income population can access nutritious and balanced meals for only one Real dollar. Amongst the donors is an organic farm that regularly gives fresh food into the food collection. The subsidized restaurant is locally known as the 'best organic restaurant in town'.

A recent trend in rich countries is the conversion of food waste into energy. The UK produces over 17 million tons of food waste every year, of which one third comes from large-scale food manufacturers. Processes have now been developed generating renewable energy from this waste. The Australian government announced the support of 1 million Australian dollars under the Renewable Energy Commercialisation Program, to design and construct 'a state-of-the-art facility that will recycle up to 82,000 tons per year of industrial and commercial food and other biomass wastes' in Sydney (Government of Australia 2005). It is praised as *food biomass to green energy* solution, without mentioning the waste of energy that occurs by not consuming the food products. Industrial agricultural production heavily relies on the use of fossil fuels. It seems just another short-sighted technological fix that does not scrutinize current food systems and wasteful production/consumption habits.

Waste Generation and Environmental Health

'The causal links between environmental change and human health are complex because often they are indirect, displaced in space and time and depend on a number of modifying forces' (WHO 2005, 2). Although most of us generate waste on a daily basis, we are unaware of environmental health issues related to the consequences of disposal and management of our waste. People do not feel responsible for the environmental impact of discarded products or materials once those items are classified as garbage.

Waste minimization, adequate disposal methods, and avoidance of littering need to be enforced and regulated. The most adequate management options vary from waste minimization (environmental education, incentives and fees) to waste diversion (reuse, recycling and composting). The best alternative is still waste minimisation, through *voluntary simplicity*, reuse, recycling, and composting.

Currently, still most waste is deposited at landfills, with sanitary landfills prevailing in the North and uncontrolled garbage dumps in the South. Landfills create a number of environmental problems and costs. Besides using up space, they release carbon dioxide and methane gas, which contribute to the greenhouse effect. In India, landfill emissions were the third largest source of greenhouse gases in the country in 1997, adding to global warming (Gupta et al. 1998). Garbage dumps, particularly if uncontrolled, are considered an environmental hazard due to gas emissions and toxic leaching often contaminating drinking water sources. Space is expensive in large cities and is often considered a limiting factor for the municipality. As a consequence, garbage is transported over large distances and is dumped in less populated regions and locations with less strict environmental regulations. Household waste is regularly moved over 100 km before it is deposited in landfills.

Incinerating waste is a common practice in high-income countries. More recently, generating energy from waste incineration is becoming an attractive concept. In Germany, this quick and apparently uncomplicated solution has stimulated some cities to occasionally make use of their newly acquired incinerators to burn household waste already separated for recycling. The toxic ashes and air pollution as a consequence of the burning process and its by-products cause health impacts. Incineration of plastics (for example, Polyvinyl Chloride – PVC and Polyethylene Terephthalate – PET) releases dioxins, furanes, heavy metals and other toxins, which are linked to the development of cancer and are known to damage the human immune system. What if developing countries that don't have the financial resources to buy the latest technology opt for outdated and risk prone incineration facilities? In reality incineration and landfilling are a tremendous waste of resources, because valuable materials are not reintroduced into the production cycle. Furthermore, neither of them are labour-intensive processes and hence do not contribute to the creation of employment. Particularly incineration rather eliminates jobs from the recycling sector. With increasing unemployment rates we need to consider these social factors in public decision-making.

Incineration, landfilling, but also lack of waste collection threatens human and environmental health. Decisions about where to place new facilities and how to treat solid waste often generate local conflicts because usually nobody wants to have such hazardous amenities in their proximity. There is a visible correlation between the distribution of hazardous activities and average income distribution. Political ecology, specifically the environmental justice movement, considers these issues.

Environmental Justice

The distribution of human induced environmental problems closely follows social-economic characterizations resulting in visible spatial patterns, differentiating spaces of poverty and degradation with those of wealth and higher environmental quality. The environmental justice movement (Bullard 1994) identifies the mechanisms that generate inequalities. It focuses on the ecological conditions of human economies and analyses the specific political and economic processes involved. Mapping the distribution of inequalities helps highlight the geographic and social differences of environmental impacts as well as the absence of resources. Bullard (1994), Pulido (1996), Markham and Rufa (1997) among several other authors, have worked on the role of race, class and poverty levels in determining waste site locations. Landscapes of waste, such as landfills, dumpsters, sewage treatment plants, recycling centres, or littered spaces in the urban fringe, are often located in spaces that are physically and perceptually marginal.

Demonstrating the spatiality of environmental equity also contributes to the understanding of the processes that lead to social exclusion. Based on the principle that no action should disproportionately disadvantage any particular social group, the social policy debate engages with visualizing inequity and injustice and with strengthening the quest for public participation. Economic exclusion can generate environmental impacts, such as deforestation and resource overexploitation in rural areas and irregular land occupation and sewage contamination in protected watersheds of urban areas. Furthermore, the distribution of these impacts can disproportionately impact on poor people's lives, which is a social injustice.

Waste management impacts differently on different social groups. The classic example is the placement of incinerators and landfills near low-income neighbourhoods, a situation found in many cities in North America and in many poor countries. The choice of location does not always reflect a deliberate planning decision, but often it is the consequence of urban development. Poor housing areas expand into locations of least real estate value, and consequently are often next to landfills, irregular waste deposits, and incinerators.

Environmental justice provides a framework that assists in the analysis of the factors that promote inequality manifested in social exclusion. The concept also uncovers the problem of excluded spaces. The periphery, particularly in large urban centres in the developing world, is the location where the unwanted is discarded and where environmental protection is least enforced. Irregular dumping of solid and liquid waste, hazardous industrial waste and harmful, polluting production frequently occur in the urban fringe. During the early 1990s the city administration of São Paulo had considered the installation of a solid waste incineration facility near a low and middle-income housing area, next to one of the two drinking water reservoirs of the city, Lake Billings. Popular mobilisation supported by the local branch of the NGO Greenpeace and other NGOs such as the Instituto Polis in São Paulo have successfully reacted in opposition to this initiative.

Environmental justice is a tool to analyse access and distribution of resources. This tool can be used to evaluate institutional conditions and processes for the enhancement of individual capacities. In her studies on the dynamics of social movements, Iris Young (1990) incorporates the concept of power into the analysis of place and community in order to understand the structural processes of oppression. The understanding of how power works in such situations is important to fully grasp the spatial and structural processes of urban poverty and the emergence of new social movements.

Growing general health awareness and community mobilisation towards environmental health issues has created significant opposition to waste management facilities, including recycling centres. People do not want to live next to visible health risk factors and therefore oppose facilities that bear a potential threat. In Brazil the homeless movement is an important player in the struggle for public housing, basic infrastructure and basic public services. More recently, leaders from the recycling sector throughout the country have also begun to organise, and many of the new co-operatives and groups that have emerged since are now part of this national movement (see Chapter 4).

Cultural Values of Waste

Political ecology recognizes theories that discuss cultural values and meanings attached to things. Valuation is a subjective, cultural and contextual phenomenon. Theories of value are crucial to environmental education (Koponen 2002). From this perspective we can understand that waste is also a culturally defined construct: it has different values or no values attached to it according to the cultural and social context. For example, with recycling, material previously defined as useless waste changes into a valuable resource in the commodity chain (Koponen 2002). Conflicts arise over these changes in the perception of value. Informal recyclers in the North and in the South – who had always seen waste as valuable – are now threatened by public or private waste management operations that have discovered the value in these resources and claim ownership over household waste. Understanding the role culture plays in the construction of value and recognizing the possible societal implications is particularly relevant in comprehending the notion of waste as a resource and in predicting and preventing possible conflicts over this resource. Furthermore, it is important to understand the specific local cultural context in order to be able to tackle waste reduction through environmental education.

Irregular garbage dumping is a growing concern in developing countries, where a significant number of households do not have access to adequate, basic infrastructure and services (sewage and waste collection, drinking water, electricity, street cleaning, and so on). Often municipal budgets are insufficient to cope with rapid population growth and increasing costs for waste collection and disposal. Irregular housing conditions and urban squatting are widespread. According to a United Nations prognosis, half the population of most Asian cities are now living in slums or squatter settlements. In some cities in Africa up to 90 per cent of the urban population lives under inadequate and risky conditions. In Dar Es Salaam, for example, 70 per cent of city residents live in unplanned areas, most of which are not regularly serviced with basic infrastructure and solid waste collection (Halla and Majani 1999). Chapter 3 will provide in-depth information on the precarious housing conditions of a significant part of the world's population.

Waste that is not collected produces expensive environmental health impacts. It contaminates water and intensifies the effects of flooding, of slope instability, and of the propagation of insects, rodents, and fungus, which can transmit infectious diseases. Many households in poor living conditions regularly burn their uncollected waste, which further adds to air contamination. Inadequate open dumping is a common problem in many developing countries often due to a lack of alternatives, lack of funds, and a general absence of environmental awareness. Dumping in fragile environments like mangroves, dunes, water catchments and floodplains, poses additional long-term threats to human health, which are sometimes irreparable.

Waste dumping is a negligent, careless act. Dumping in the periphery by locals and outsiders – particularly small, inert, waste management and transportation businesses – further reinforces the condition of marginalization experienced by the people living in these areas (see Illustration 2.1). These acts of disregard towards the environment underscore the lifelong exclusion of this population. It impacts on the aesthetics of the place, hindering the creation of a *sense of place* and place identity of its transient population. Oppression and domination, in the sense of perpetuating inequities, has created these excluded spaces, where everything seems under risk,

Illustration 2.1   Irregular waste deposit in a squatter settlement next to the border between São Paulo and Diadema

transitional and non-committing. Iris Marion Young's politics of difference (Young 1990) helps de-mask these conditions by dismantling and reforming the structures, processes, concepts and categories that sustain the current politics and policies.

The Global Rise in Recycling: An Option?

Increased packaging, changing lifestyle and consumer attitudes are responsible for the increase in the percentage of recyclable waste. Although recycling contributes to waste recovery, it must be reiterated that it is not the optimum solution, and unfortunately in rich countries the emphasis on recycling draws attention away from root causes. Citizens feel that they have done their part and don't question consumption patterns and lifestyles. Furthermore, often recycling programmes themselves induce unsustainable behaviours, such as driving to bottle banks and drop-off spots, transporting recyclables over large distances, or using plastic bags to collect and deposit recyclables.

Packaging waste is the main component of domestic waste. Where recycling programmes are widely accepted, packaging waste has in most cases not significantly diminished (OECD 2002). Today approximately 40 per cent of the paper, cardboard and glass wasted are recovered from the domestic waste stream in Europe and North America. In Germany, Austria and Switzerland, bottle banks now recover more than 75 per cent of the glass bottles consumed. In 2001, the UK implemented new policies, which require the recovery of 50 to 65 per cent of packaging from household waste, including the recycling of at least 15 per cent of all plastics in use (McDonald and Ball 1998). In 2002, 22.7 per cent of plastic packaging waste was recovered in the UK (DEFRA 2004). As in Germany, the Irish government has successfully diminished the number of disposable polyethylene bags through the increase from 15 cents to 22 cents tax on every plastic shopping bag. While the UK Government has no plans to introduce a tax on plastic bags at present, the British retail sector in February 2007, announced a voluntary agreement to reduce the overall environmental impact of carrier bags by 25 per cent by the end of 2008 (DEFRA 2007). Currently, about 8 billion plastic bags (134 bags/year/person) are wasted every year as garbage. Shopping bags are just a fraction of our plastic packaging waste. In spite of the environmental impacts of the toxicity and lack of biodegradability of plastic waste, the use of plastic bags continues to rise – more so in developing countries (UNEP 2002). Change is possible, as demonstrated by Co-op supermarkets in Italy, where only biodegradable plastic shopping bags are provided. It needs to be noted though that biodegradable plastic bags also use up energy and release pollutants during the process of degradation. Why not reuse fabric shopping bags?

According to the Brazilian plastic packaging association ABPET (Associação Brasileira de Embalagens PET), approximately 70 per cent of all soft drinks in Brazil are packaged in PET (Polyethylene Terephthalate) bottles (Partido Verde de Minas Gerais 2004). In 2001, the production was around 360,000 tons and the consumption within Brazil was 270,000 tons, of which only 89,000 tons were recycled (or 25 per cent of the total production and 33 per cent of the total consumption) (Partido Verde de Minas Gerais 2004).

Corporate Interest in Recycling

The business sector in Brazil has a vested interest in accessing the wealth embodied in household waste. The initiative CEMPRE (Compromisso Empresarial para Reciclagem) promotes recycling by providing specific credit lines, incentives, and technical support to businesses interested in the recycling sector. Multinational corporations like Rhodia-Ster (Mossi and Grisolfi Group), for instance, not only produce PET drinking bottles, but are also involved in the business of recycling them. The number of recycling firms in the formal sector, in Brazil, has increased from 95 in 1996, 232 in 2000 and 740 establishments in 2005, with 15,917 persons directly involved in the formal recycling sector by the end of 2005 (IBGE 2007). This reveals a significant contribution of the sector to the national economy.

Not only has recycling become attractive to the large-scale business sector, but waste management has become an attractive economic profit sector. As a consequence, large multinational enterprises (such as Onyx, Vivendi, Vesta) are increasingly taking over waste management when municipal services are privatized (Bartone et al. 1991, Kaseva and Mbuligwe 2005; Lee 1997; Ogu 2000; Post et al. 2003). A proposal receiving more attention within governments, business and academics is the *waste to energy* concept, whereby waste (often including the recyclable portion as well) will generate energy through incineration, as was previously discussed in the context of transforming food waste into energy. Usually the discussion omits the fact that incineration also wastes resources, and that this proposition jeopardizes the livelihood of the many people that are already making a living from recovering waste. It also does not consider the possible health threats from incineration, as discussed earlier. The case studies in the following chapters will consider the possible impacts of waste management on informal and organized recyclers in poor countries, and will showcase *third ways*.

Informal Recycling a Survival Option

Historically, recovering recyclable and reusable material from garbage has always been a livelihood option everywhere. Until the end of the nineteenth century most cities in North America lacked municipal waste collection, and street and landfill scavenging were common activities, particularly among immigrants and poor people (Ross 1996). Medina describes the situation in New York, where Italian immigrants frequently searched through the waste transports before they reached the dumping sites (Medina 2001). Observing this particular case is interesting because it shows how public policies have changed following an evolution in the perception of waste from a valueless to a valuable resource. Until 1878, the city paid for the services and allowed the scavengers to keep what they had salvaged. From 1878 to 1882, they were still allowed to recover materials but were not paid any more. After that the local authorities charged a flat rate to scavengers for accessing the city's refuse (Medina 2001, 232). Similar situations are described for other places and countries, where recycling has become a widespread income generating activity (Perry 1978; Ross 1996). Until the 1950s, in many cities throughout North America, the poor collected materials from the dumps. This practice was banned due to sanitary considerations and to prevent potential liability suits from scavengers. Estimates indicate that there were approximately 8,000 scavengers in New York City during the mid 1990s (Rendelman and Feldstein 1997). Despite the prevalence of this activity in North America, there is limited information available on the extent of this population, on its significance towards waste management, and the role government plays in supporting or inhibiting this activity.

In poor countries recycling is widely practised, albeit mainly on an informal basis or involving organized co-operatives and associations (see Chapter 4). It has been extensively described as a survival strategy in Asia (Beall 1997; Medina 2000; Sarkar 2003), in Latin America (Ali 1999; Ojeda-Benitez et al. 2002; Berthier 2003), in Africa (Kaseva 1996; Adeyemi 2001; Fahmi 2004, and in North America (Rendleman and Feldstein 1997; Medina 2001).

Recycling is a very diverse sector with obvious social, cultural, and economic differences changing from place to place. There are many ways in which recyclables are separated from garbage. Often the activity involves everyone: children and the elderly, women and men. Sometimes entire families live on garbage dumps or scavenge in the streets. Different modes of transportation and various organizational structures are used, and demonstrate the creativity and resourcefulness of the people involved. Hand-pulled carts, self-crafted out of discarded materials such as wood and cardboard, are most common. Often even old refrigerators get transformed into pushcarts. For most recyclers, recycling is the main income source (see Illustration 2.2).

Illustration 2.2   Informal recycler in São Leopoldo, Brazil

Some recyclers work independently; others are part of organized groups, co-operatives, associations or governmental waste management programmes. This sector is also referred to as popular economy or solidarity economy (Barros 2003; Magera 2003; Singer 2003). The recyclers conform to territorial divisions depending on factors such as participation in an association or co-operative, power ranking within the universe of recyclers, and length of time they have been performing the activity. As they are not only socially excluded but also stigmatized for their work, they are highly vulnerable. Often it is entirely the informal sector which does the collection, disposal and sorting of solid waste, without any financial support or technological resources from the local government (Medina 2000; Van Horen 2004). Significant change is recently happening in Brazil, with some municipalities and the federal government setting the precedence with giving recycling co-ops the priority in waste diversion contracts. The law (Decreto) No. 48.799, signed on 10 October 2007 has now made this policy official in the city of São Paulo (Diario Oficial 2007). On 16 October 2007, the Brazilian congress also received the law proposal 1991/2007, as part of the Federal solid waste management law, which would guarantee recycling co-operatives and associations the primary right to be contracted in formal resource recovery programmes (Instituto Akatu pelo Consumo Consciente 2007).

Urban Growth and Waste: The Case of São Paulo, Brazil

The municipality of São Paulo is located within the metropolitan region of São Paulo (MRSP), an urban agglomeration of approximately 18 million dwellers. The city's population increased from 6 million in 1970 to over 10.4 million in 2000, and it is the core of the world's third largest metropolitan agglomeration. Arguably, São Paulo is a global city. It is the most important finance and service centre in South America that is surrounded by other major industrial production centres. It is also surrounded by poverty, visible in the widespread periphery, with squatter settlements and *favelas*. Over the past decades the southern fringe of the city has developed into spaces of social and environmental exclusion. The following map (Figure 2.2) shows the high incidence of irregular, open dumping of solid and liquid waste in the Billings watershed, at the southern fringe of São Paulo. The close localization of three major landfills within this catchment is another obvious fact reiterating environmental exclusion.

Although the growth rate has declined significantly since 1970 – the boom period of the rural exodus – São Paulo still attracts landless and homeless people displaced from the countryside. In the city, most jobs available to them are in the informal service sector. Automation in industry is excluding more workers from employment and the number of street vendors and scavengers has visibly increased over the past fifteen years. During the 1990s the number of people dependent on informal activities has grown 34 per cent in São Paulo. In 1999, one out of people in the labour force was working in the informal sector (Martins and Dombrovski 2001).

There are many obvious social and environmental impacts from the rapid urbanisation in São Paulo. The expansion of basic infrastructure and public services has not accompanied the fast pace of the city, which leaves part of the population unattended. Those living under irregular conditions and on the urban fringe usually

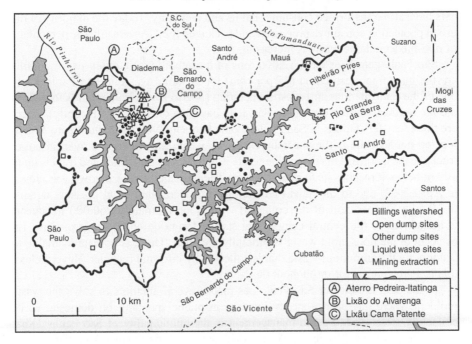

Figure 2.2   The Billings watershed in southern São Paulo and irregular waste deposits

Sources: Capobianco, J.P.R. and Whateley, M. (2002) and *Billings 2000*, (São Paulo: Instituto Socioambiental).

Cartography: Ole J. Heggen.

improvise for their housing, water and electricity, often at the expense of the environment. Untreated sewage discharge and waste disposal are major problems in São Paulo, especially in the outskirts. During the early 1990s the city rapidly expanded southwards into the drinking water catchments of Lake Billings and Lake Guarapiranga. Serious conflicts arouse due to the undermining of the city's water supply and water shortages were becoming more frequent during the dry season.

As in other countries, patterns of consumption and waste disposal have changed during the past decade. Today, the metropolitan region of São Paulo produces an average of 20,150 tons of waste/day (IBGE 2002). On a per person basis waste generation has grown from 0.89 kg/person/day in 1991, to 0.99 kg/person/day in 1994, to a high of 1.16 kg/person/day in 2001 (IBGE 2002). Only 54 per cent of the final destination of the solid waste generated by this region is considered adequate, while the rest is considered inadequate 10 per cent are considered under high risk conditions (JICA 2004).

Following the global trend, packaging waste has also grown in São Paulo, particularly one-way packaging. Most beverages are now bottled in PET (Polyethylene Terephthalate), aluminium containers, or *Tetra Pack* (plastic- and aluminium-foiled cartons). The availability of fast food and takeaway and the increased use of private cars have also transformed consumer habits and mentality. Today the majority of the urban population consumes processed and packaged food. As a result, the percentage of non-biodegradable waste is increasing.

São Paulo generates circa 12,000 tons of household waste every day. Landfills are the main final destination (77 per cent) for the solid waste in São Paulo. Only one of the controlled landfills is still operating (Aterro Bandeirantes), it is located in the far Northeast of the city (Sítio São João has been closed down recently). There are three transfer stations in São Paulo (Ponte Pequena, Vergueiro and Santo Amaro located north of the Billings catchment) from where the waste is moved in larger trucks to the landfill. This means the waste is handled several times and travels up to 60 km before it reaches its final destination. The only incinerator in town was closed in April 2002, due to inefficient technology and increasing local pressure against the facility. 21 per cent of the collected waste (approximately 260,000 tons/year) is transferred to the central composting station Usina de compostagem. Only 1.6 per cent of the household waste is officially recycled. There is no data available to account for the amount of waste that is collected by informal recycler. Nevertheless, this is a growing sector throughout the city.

In the past, the city of São Paulo has made several attempts to solve the waste problem. There has been a steady increase in public spending on domestic waste collection, street cleaning and management in the municipality of São Paulo. Other expenditures of the Municipal Waste Management Department (Departamento de Limpeza Urbana) include running and maintenance costs of waste treatment plants; expansion and implementation of incinerators, landfills and composting units; general administration; public information and education services; household waste recycling; and collection and final disposal of hospital waste. Most of these services, including waste collection, have been subcontracted to private firms by the municipality. Between 1988 and 1999, out of the total waste management costs, 24 per cent were spent on administration, 35 per cent on collection and 41 per cent on cleaning of public spaces.

In São Paulo the responsibility for solid waste services lies with the metropolitan authority LIMPURB, which is a municipal department that subcontracts all services. Until 2004 three large firms (VEGA-SOPAVE, CAVO and ENTERPA) were contracted for collection, street sweeping, transfer, landfills, composting and incineration plants (Figure 2.3). Since 2004, only two corporations are responsible for waste amangement: Ecourbis in the south and Loga Logistica Ambiental S.A. in the north. As in many other cities in Brazil, corruption and nepotism have frequently undermined the existing public sector, affecting the quality of the services. In early 2000, an inquiry discovered the misappropriation of large sums from the city's waste management budget, involving the mayor of São Paulo (Celso Pitta) as well as large firms contracted by LIMPURB. The administration was accused of irregular overspending in waste management, while at the same time many complaints were voiced about the negligence in garbage collection and street cleaning, particularly in the periphery. Between 1988 and 1999, the municipality had spent approximately Reais $4.3 billion (US$2.4 billion) on municipal waste management (of which a significant proportion was mismanaged). Only a negligible amount of this budget was used for recycling, composting and other alternative waste management programmes. In May 2000, as a result of this and other scandals, the mayor was impeached.

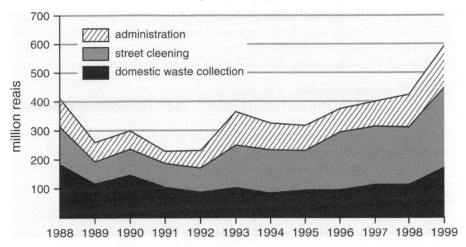

Figure 2.3   Increase in public spending on waste management in the municipality of São Paulo (1988–1999)

Source: São Paulo. *Sistema de execução orçamentária – SEO* (Assessoria Econômica – Gabinete Vereador Adriano Diogo).

Cartography: Ole J. Heggen and Jutta Gutberlet.

In 2003, the municipality of São Paulo launched a recycling programme, Coleta Seletiva Solidária, foreseeing the construction of 30 classification centres, as well as capacity building for the groups involved in the separation of the recyclables. By 2005, out of the 15,000 tons of solid waste generated daily by the city 1,300 tons were recycled. The municipal service collected 70 tons and the informal organized recyclers recovered 1,230 tons per day (Instituto Polis 2006). The material was separated at the 15 classification centres, involving approximately 700 recyclers, whose income had increased to circa Reais $500 with the programme. Furthermore, the official collection of recyclables had increased 230 per cent since 2003 (Instituto Polis 2006). However, with the change in government at the beginning of 2005, as happens so frequently in Brazil, the existing recycling programme was interrupted and the expansion plans of building new recycling facilities was abandoned. The organized recycling groups had to renegotiate with the new municipal government for the continuation of the programme. It is very common that the lack of continuity during and after government transitions affects ongoing social programmes and policies, as was the case with this recycling programme.

Official budget data for the city's 2008 projection hint a significant increase in promoting selective waste collection and recycling. The city has forecasted to spend almost Reais $50 Million in 2008 on the expansion and infrastructure maintenance for selective waste collection and separation implemented through organized recycling co-operatives and associations. Another million Reais dollars will be spend on environmental education and Reais $950,000 are projected for capacity building activities with the recyclers. These developments are in line with the federal government, which is supportive to inclusive resource recovery.

In Chapter 4 the role and extent of the informal and organized recycling sector will be discussed, providing a local, community perspective from the outskirts of São Paulo. The previously mentioned recent innovations in the public policies at the local and federal level are promising. Time will show whether they can make a difference.

Conclusion: Contributions to Urban Inclusion and Environmental Health

There is a strong relationship between rapid urban growth and global environmental change. With the urban lifestyle comes an ever-growing generation of solid waste, which in many large cities is already undermining human security. Most domestic waste is deposited in sanitary landfills or, in poor countries, at irregular dumping sites, often frequented by scavengers or *catadores* recovering the recyclables to maintain their livelihoods. The lack of space for new landfills is an imminent problem for municipalities all over the world, often causing the waste problem to be transferred to more distant places.

There is no doubt that, on an individual level, voluntary simplicity and responsible consumption would help reduce the waste of resources, the waste of environment and landscapes and the waste of people. In order to make a significant difference, industrial consumption and public procurement need to be dealt with. Government and business-to-business procurement account for a large part of overall consumption. An important mechanism for environmental improvement is to change all scales of industrial supply chains. Policies can directly change consumption outcomes. We have seen that solid waste generation is a worldwide problem and affects people in all parts of the world. A revision of consumption patterns and a reorientation towards voluntary simplicity and responsible consumption on a global scale may be addressed following these propositions:

- Government needs to provide incentives for resources conserving consumption patterns, rather than rely on the environmental consciousness of consumers Products which are more environmentally friendly (for example, that require little energy in its production) and socially just (fair trade) need to be cheaper to stimulate the consumption of these items.
- Policy-makers need to focus on areas where consumers can make a difference: for example, in the areas of packaging, transportation, and energy.
- Measures to encourage environmental and social conscious consumption need to act across the very different domains of industrial, public and private consumption. Citizens need to have the opportunity to be 'green' and 'fair' while acting in these different spheres.
- Waste management options need to take into consideration not only environmental facets but principally social economic aspects such as income generation.

Recycling is an alternative form of waste management; but, more importantly, it can also be a strategy to diminish unemployment. There is a unique opportunity to tackle human security issues with adequate and inclusive waste management strategies. Until today the potential for resource recovery has been neglected. Only a few cities embrace innovative alternatives, where social and environmental policies engage in reduction, reutilization, and recycling of materials, also providing a chance for the poorest sector of the population to participate.

There are serious social and environmental challenges inherently linked to the disposable society that need to be addressed. Cities are the places where poverty is most concentrated in Latin America. It is very likely that over the next years this will also be true of Asia and Africa. On a global level, we are facing environmental health predicaments due to consumption and generation of waste. Littering, dumping, landfilling, and burning solid waste cause severe impacts on the urban populace and environment in the North and the South. Groundwater and surface water, the most important basic resources, are under threat from waste contamination. Particularly solid waste landfill sites pose severe health risks due to landfill leachate. Widely practiced discharge of untreated sewage and industrial effluents, have severely compromised urban surface waters. In addition, littering, so obvious and common, have already contaminated the rivers and lakes in most cities. A true commitment to urban sustainability means tackling human security, focusing on urban poverty.

Human security can be enhanced. It is important to support individuals and communities in their struggle to end threats to their livelihood. In the short term, strategies need to be found to mitigate these threats; in the long term, communities must be made better able to adapt to these threats, resulting in more sustainable livelihoods. The concept of human security recognizes that supporting effective local solutions and answers from within the community is the most productive and preventive approach. It highlights the potential for a people-centred approach, with individuals and communities actively participating in pursuing options that lead to human security, as outlined in the following two recommendations:

1. Addressing human security with the generation of *sustainable* employment (which means employment with a low footprint). Promoting labour-intensive practices through recycling, re-use or prevention (environmental education and awareness building campaigns are effective ways) is a concrete example that works. Health risks need to be eliminated through the introduction of sound technology. Adopting political accountability and participatory, integrated decision-making strategies are more likely to result in transparency of human security or insecurity issues.

2. Minimising the impacts of global environmental change. Promoting less resource-intensive lifestyles and values, criticizing the apparent right we seem to seize to waste resources, and promoting values that respect the environment and value our bioregion can all help reduce environmental impacts. Stimulating habits of reduction, re-use, and recycling (for example with Eco-taxes), supporting the development of biodegradable products and cutting over-consumption are concrete possibilities whose results have already been tested and verified in the context of different places.

Solving the waste problem requires political solutions. The case of São Paulo highlights important changes that can be brought about with progressive government programmes. The example also illuminates the detrimental bottlenecks created by lack of governmental commitment to ongoing social programmes, particularly during phases of political change. It also emphasises the difficulties encountered when addressing social and environmental problems from an integrated and participatory perspective.

Ultimately the desired changes in production and consumption as well as the increase in human security and sustainability will depend on political change, in particular on the political will to change. Technical knowledge and financial resources are not the limiting factors in promoting urban sustainability. Human and financial resources from the public and private sector must be efficiently used to build more sustainable cities. The main problems are political, and require a perspective that fosters horizontal and vertical co-operation among government agencies and among the community, the public and the private sector. A strongly committed approach to socially and environmentally sound urban development is essential in order to diminish the threats of deprivation and over-consumption.

References

Adeyemi, A.S., Olorunfemi, J.F. and Adewoye, T.O. (2001), 'Waste Scavenging in Third World Cities: Case Study in Ilorin, Nigeria', *The Environmentalist* 21, 93–6.

Ahmed, S.A. and Ali, M. (2004), 'Partnership for solid waste management in developing countries: linking theories to realities', *Habitat International* 28, 467–79.

Ali, M. (1999), 'The informal sector: What is it worth?', *Waterlines* 17(3), 10–12.

Alkire, S. (2003), *A Conceptual Framework for Human Security*, CRISE Working Paper 2 (Oxford: CRISE).

Barros, C.J. (2003), 'Rede Solidária: Universidades atuam na formação e capacitação de cooperativas', *Problemas Brasileiros* July/August, 22–31.

Bartone, C.R., Leite, L., Triche, T. and Schertenleib, R. (1991), 'Private Sector Participation in Municipal Solid Waste Service experiences in Latin America', *Waste Management and Research* 9, 495–509.

Baud, I., Grafakos, S., Hordijk, M. and Post, J. (2001), 'Quality of life and alliances in solid waste management. Contributions to urban sustainable development', *Cities* 18:1, 3–12.

Beall, J. (1997), 'Thoughts on Poverty from a South Asian Rubbish Dump', *Institute of Development Studies* 28(3), 73–89.

Berthier, H.C. (2003), 'Garbage, work and society', *Resources Conservation and Recycling* 39, 193–210.

Bilitewski, B., Härdtle, G., Marek, K., Weissbach, A. and Boedicker, H. (1994), *Waste Management* (Berlin: Springer).

Britannia Community Services Centre (2006), 'Connecting the dots. Mapping local food resources in Grandview/Woodlands', A project of Grandview/Woodlands Food Connection.

Bullard, R.D. (1994), *Dumping in Dixie: Race, Class, and Environmental Quality* (Boulder, CO: Westview Press).

Burgess, J., Bedford, T., Hobson, K., Davies, G. and Harrison, C. (2003), '(Un)sustainable consumption', in Berkhout, F., Leach, M. and Scoones, I. (eds), *Negotiating Environmental Change: New Perspectives from the Social Sciences* (Cheltenham: Edward Elgar Publishing Limited).

Chung, S. and Poo, C. (1998), 'A comparison of waste management in Guangzhou and Hong Kong', *Resources, Conservation and Recycling* 22, 203–216.

Cotton, A., Snel, M. and Ali, M. (1999), 'The challenges ahead – solid waste management in the next millennium', *Waterlines* 17(3), 2–5.

CRD (Capital Region District) (2007), 'Composting & Organics Recycling', [website] http://www.crd.bc.ca/waste/organics/index.htm, accessed 9 September 2007.

DEFRA (Department for Environment, Food and Rural Affairs) (2004), Information Bulletin (47) 04, London, [website] http://www.defra.gov.uk accessed 6 September 2007.

—— (2007), [website], updated 30 March 2007, http://www.defra.gov.uk/environment/localenv/litter/plasticbags/index.htm, accessed 12 November 2007.

Diário Oficial (2007), LEI No. 14.512, DE 9 DE OUTUBRO DE 2007 (Projeto de Lei No. 353/06, do Vereador Toninho Paiva – PR), Cidade de São Paulo, 52 (189), 10 October 2007.

European Union (2006), 'The European Parliament, Directive of the European Parliament and of the Council on Waste', [website], updated 14 July 2007, http://www.wastexchange.co.uk/documenti/europeanorm/st03652-re01.en05.pdf, accessed 9 September 2007.

Fahmi, W.S. (2005), 'The impact of privatization of solid waste management on the Zabaleen garbage collectors of Cairo', *Environment and Urbanization* 17:2, 155–70.

Fehr, M., de Castro, M.S.M.V. and Calcado M.d R. (2000), 'A practical solution to the problem of household waste management in Brazil', *Resources, Conservation and Recycling* 30:3, 245–57.

Forkes, J. (2007), 'Nitrogen balance for the urban food metabolism of Toronto, Canada', *Resources Conservation and Recycling* 52(1), 74–94.

GECHS (Global Environmental Change and Human Security) (2007), [website], updated 14 April 2007, http://www.gechs.org/human-security/, accessed 1 September 2007.

Government of Australia (2005), 'Food biomass to green energy', Australian Greenhouse Office, Department of the Environment and Water Resources, [website], updated 11 July 2007, http://www.greenhouse.gov.au/renewable/recp/biomass/seven.html, accessed 12 November 2007.

Gupta, S., Mohan, K., Prasad, R., Gupta, S. and Kansal, A. (1998), 'Solid waste management in India: options and opportunities', *Resources, Conservation and Recycling* 24, 137–54.

Halla, F. and Majani, B. (1999), 'Innovative ways for solid waste management in Dar Es Salaam: Toward stakeholder partnerships', *Habitat International* 23(3), 351–61.

Hasan, S. and Khan, M.A. (1999), 'Community-based environmental management in a megacity', *Cities* 16(2), 103–110.

IBGE (Instituto Brasileiro de Geografia e Estatística) (2007, 2002), Industrial census data: Pesquisa Industrial Annual 2007, 2002, [website] http://www.ibge.gov.br, accessed 12 April 2007.

Instituto Akatu pelo consumo consciente (2007), 'Chega ao Congresso o projeto de lei que trata do lixo', 16 October 2007, [website] http://200.169.96.83:8081/akatu, accessed 11 November 2007.

Instituto Polis (2006), 'São Paulo e os catadores de materiais recicláveis', *Especial Dossiê* (São Paulo: Instituto Polis).

Jayne, M. (2006), *Cities and Consumption* (London: Routledge).

JICA (2004), 'Avaliação da situação dos resíduos sólidos no brasil, no estado de São Paulo, na região metropolitana de São Paulo e no município de São Paulo. São Paulo', [website] http://ein.jica.org.br/downloads/resíduos%20sólidos%20-%20São%20Paulo.pdf, accessed 12 April 2007.

Kaseva, M.E. and Mbuligwe, S.E. (2005), 'Appraisal of solid waste collection following private sector involvement in Dar Es Salaam city, Tanzania' *Habitat International* 29, 353–66.

Köberlein, M. (2003), *Living from Waste: Livelihoods of the Actors Involved in Delhi's Informal Waste Recycling Economy* (Saarbrücken: Verlag für Entwicklungspolitik).

Koponen, T.M. (2002), 'Commodities in Action: Measuring Embeddedness and Imposing Values', *Sociological Review* 50(4), 543–69.

Lee, Y.-S.F. (1997), 'The privatization of solid waste infrastructure and services in Asia', *TWPR* 19:2, 139–61.

Magera, M. (2003), *Os Empresarios do Lixo. Um Paradoxo da Modernidade* (Campinas: Editora Atomo).

Markham, W.T. and Rufa, E. (1997), 'Class, race, and the disposal of urban waste locations of landfills, incinerators, and sewage treatment plants', *Sociological Spectrum* 17, 235–48.

Martins, R. and Dombrowski, O. (2001), 'Mapa do trabalho informal da cidade de São Paulo', in Jakobsen, K., Martins, R. and Dobrowski, O. (Org.) *Mapa do trabalho informal* (São Paulo: Fundação Perseu Abramo).

McDonald, S. and Ball, R. (1998), 'Public participation in plastics recycling schemes', *Resources, Conservation and Recycling* 22, 123–41.

Medina, M. (2001), 'Scavenging in America: Back to the Future?', *Resources, Conservation and Recycling* 31, 229–40.

Mee, N., Clewes, D., Phillips, P.S. and Read, A.D. (2004), 'Effective implementation of a marketing communications strategy for kerbside recycling: A case study from Rushcliffe, UK', *Resources, Conservation and Recycling* 42(1), 1–26.

Mollison, B. (1990), *Permaculture: A Practical Guide for a Sustainable Future* (Washington: Island Press).

OECD/Eurostat (2006), EIONET: 'European Topic Centre', [website], updated 14 July 2007, http://waste.eionet.europa.eu/definitions/waste, accessed 1 September 2007.

Ogu, V.I. (2000), 'Private sector participation and municipal waste management in Benin City, Nigeria', *Environment and Urbanization* 12(2), 103–117.

Ojeda-Benitez, S., Armijo-de-Vega, C. and Ramirez-Barreto, E. (2002), 'Formal and Informal Recovery of Recyclables in Mexicali, Mexico: Handling Alternatives', *Resources, Conservation and Recycling* 34: 273–88.

Perry, S.E. (1978), *San Francisco Scavengers: Dirty Work and the Pride of Ownership* (California: University of California Press).

Pöll, G. and Schneider, F. (1993), *Returnable and Non-returnable Packaging. The Management of Waste and Resources an Eco-social Market Economy* (London: James & James).

Pongracz, E. and Pohjola, V.J. (2004), 'Re-defining waste, the concept of ownership and the role of waste management', *Resources, Conservation and Recycling* 40, 141–53.

Price, J. and Joseph, J.B. (2000), 'Demand management – A basis for waste policy: A critical review of the applicability of the waste hierarchy in terms of achieving sustainable waste management', *Sustainable Development* 8, 96–105.

Pulido, L. (1996), *Environmentalism and Economic Justice* (Tucson: University of Arizona Press).

Rendleman, N. and Feldstein, A. (1997), 'Occupational Injuries Among Urban Recyclers', *Journal of Occupational and Environmental Medicine* 39(7), 672–5.

Ross, A. (1996), 'The Lonely Hour of Scarcity', *Capitalismo Natura Socialismo* 7(3), 3–26.

Sarkar, P. (2003), 'Solid Waste Management in Delhi – A Social Vulnerability Study', Proceedings of the Third International Conference on Environmental Health, Chennai, India, Department of Geography, University of Madras and Faculty of Environmental Studies, York University.

Singer, P. (2003), 'As grandes questões do trabalho no Brasil e a economia solidária', *Proposta* 30:97, 12–16.

Statistics Canada (2000), 'Waste Management Industry Survey, 2000 – Business and Government Sectors Survey Guide Environment Accounts and Statistics Division', [website] http://www.statcan.ca/english/sdds/document/1736_D1_T1_ V1_E.pdf, accessed 1 September 2007.

UN-HABITAT (United Nations Human Settlement Programme) (2006), *The State of the World's Cities Report 2006/2007* (London: Earthscan).

UNDP (United Nations Development Programme) (1994), *Human Development Report* (Oxford: Oxford University Press).

UNEP (United Nations Environment Programme) (1999), *GEO 2000: Global Environmental Outlook*. Division of Environmental Information, Assessment and Early Warning (DEIA & EW) United Nations Environment Programme, [website] http://www.grida.no/geo2000/index.htm, accessed 12 April 2007.

UNFPA (United Nations Population Fund) (2001), *State of World Population 2001 Chapter 3: Development Levels and Environmental Impact*, [website] http://www.unfpa.org/swp/2001/english/ch03.html, accessed 12 April 2007.

——— (2004), *State of World Population 2004: Migration and Urbanization*, [website] http://www.unfpa.org/swp/2004/english/ch4/index.htm, accessed 12 April 2007.

Van Horen, B. (2004), 'Fragmented Coherence: Solid Waste Management in Colombo', *International Journal of Urban and Regional Research* 28(4), 757–73.

Warwick, H. (2001), *Cuba's Organic Revolution*, Forum for Applied Research and Public Policy, 54–8, [website] http://forum.ra.utk.edu/Archives/Summer2001/cuba.pdf, accessed 12 April 2007.

WHO (World Health Organization) (2005), *Global Environmental Change*, [website] http://www.who.int/globalchange/ecosystems/ecosystems05/en/index.html, accessed 12 April 2007.

Worldwatch Institute (2004), 'State of the World 2004: Consumption by the numbers', World Watch Institute Press Release, [website], last updated January 2004, http://www.worldwatch.org/node/1783, accessed 12 November 2007.

Young, I.M. (1990), *Justice and the Politics of Difference* (Princeton: Princeton University Press).

Chapter 3

Surviving at the Urban Frontier

Introduction: Local Development and Quality of Life in Poor Settlements

The risk-prone and unhealthy circumstances under which an increasing part of the world's population is living nowadays is unprecedented. One facet of the problem relates to the growing income disparity and unequal distribution of resources between the North and the South and among the rich and the poor in developing countries. Often bottom-up and grassroots initiatives are engaged in improving the quality of life of the poor in marginalized settlements. These proposals become important pathways towards achieving more sustainable communities.

In this chapter a case study will contextualize the general debate on urban development and livelihoods by bringing to the forefront the rise and transformation of a neighbourhood in the outskirts of São Paulo. The place is called *Pedra sobre Pedra* (stone-over-stone) and its people tell a story of vulnerability and deprivation, environmental degradation and the frustration due to the omissive behaviour of the public sector in addressing the needs in terms of housing and basic infrastructure of the poor. It is a story lived in many marginal settlements or shantytowns in developing countries. It is also about people who engage in a struggle to improve their livelihoods. In this particular case, dynamic urban development processes during the late 1980s have transformed an abandoned quarry into a landfill for inert materials and then to a rapidly expanding squatter settlement. During the late 1990s and early 2000, the municipal government implemented a programme to improve the basic infrastructure of this neighbourhood. The case study demonstrates proactive and mitigating strategies of local leaders to address the deprivation at the fringe and it illustrates the difficulties encountered in this process. Livelihood assessment is used to better understand the quality of life issues among the marginalized and poor (Convey et al. 2002). This framework is particularly useful, when it comes to look at basic human needs issues, such as access to food, shelter, work, basic infrastructure and services and so on. It organizes these assets into human, social, political, financial and natural resources capital, necessary to make a living (Convey, 2001; Bauman and Sinha, 2001). Understanding the livelihoods of the poor reveals the strategies in place to overcome difficult times and circumstances. Initially the framework was designed for the rural context (Brocklesby and Fisher, 2003; Chambers and Conway, 1992) but more recently it is also applied to the urban setting (Chant, 2004; Farrington, Tamsin and Walker, 2002; Rakodi and Lloyd-Jones 2002; Rouse and Ali 2001) and particularly to understand the assets and barriers of urban poor (Meikle, 2002).

The case study indicates that urban development needs to be re-defined in the sense of sustainable community development. Sustainable communities have an 'efficient use of urban space, minimize their consumption of essential natural capital, multiply the social capital and mobilize its citizens and the governments towards these ends' (Roseland 1998, 24). Before discussing sustainability, the different meanings of development and its purpose need to be highlighted. The central questions in this context are: how can development benefits become more equitably distributed and accessed? Where do the concepts of social cohesion, sense of place and participation fit in a new urban development framework?

It is unacceptable that today one out of three city dwellers lives under *shantytown* (or *favela*) conditions, which are defined by the UN-HABITAT as '...living in difficult social and economic conditions that manifest different forms of deprivation – material, physical, social and political' (UN-HABITAT 2006, 11). This source shows that today approximately 998 million people live in sub-standard housing conditions; vulnerable because of unsafe water supply, lacking basic infrastructure and unsatisfactory public services. The UN-HABITAT prediction that there will be a worldwide 1.4 billion poor dwellers living in *shantytowns* by 2020 is alarming. The physical expression of poor living conditions becomes manifest in lack of water, lack of sanitation and overcrowding; these conditions are captured by the term *shelter deprivation*. Another important indicator that characterizes *shelter deprivation* is *insecurity of tenure* (UN-HABITAT 2006). In Latin America and the Caribbean, the population living in precarious housing conditions has increased from circa 111 million in 1990 to 134 million in 2005, at an annual growth rate of 1.28 per cent (UN-HABITAT 2006, 16). Brazil currently houses 39 per cent of that population and compared to other countries in the region, it has a slow *favela* growth rate, with 0.34 per cent yearly increase (UN-HABITAT 2006, 30). The destitution and precariousness in these settlements and the large number of people affected make this a major concern for the government. Extreme urban inequalities continue to persist throughout the country, despite examples of recent proactive policies addressing the needs of the poor in the region. There are obvious links between the deprived living conditions in poor settlements and levels of crime and violence in the city. Excluded spaces facilitate the origin and dissemination of crime. The case portrayed here provides insights to the living conditions and the difficulties in terms of appropriate policies to tackle this human security issue.

We need to understand the driving forces behind global development – forces impelled by global financial and cultural systems. Global cultural values and habits are propagated, under the influence of large media concerns like News International and multinational corporations such as Coca Cola, Monsanto or IBM. Large metropolises have become the staging places for these global chain activities, with mass consumption and '...ceaseless flows of people, money, commodities, ideas, information and cultural influences' (Allen, Massey and Pryke 1999, 8). Understanding the prevailing global development processes allows us to grasp the forces and consequences on the local development scale. The case study will therefore contextualize our search for effective and sustainable development. Hands-on approaches have been successful in tackling some of the social, economic and environmental problems of urban livelihoods. Several of these successful cases will be presented in the hope that they might inspire proactive change in other places.

Development for Whom and with Whom?

Development is predominantly conceived from an economic perspective, equating increased wealth and economic progress, which can be translated into per capita income, Gross National Product (GNP), Gross Domestic Product (GDP) or Purchasing Power Parity (PPP). It is also perceived as change – or progress – from traditional to modern technologies, from agricultural to industrial or post-industrial societies, from small-scale to large-scale, local to global change. This positivist approach, based on the Cartesian understanding of fragmented, and primarily quantitative – hence partial – perceptions, has ruled the development debate for many decades. Social theory contributes to unveiling the underlying dominant paradigms of the development approaches, from classical to contemporary theories. Liberal growth-oriented thinking as defended under Rostow's modernization theory has deeply impacted the South. Development, narrowly defined as economic growth, sustained by our *Westernized* worldviews and the predominance of technical scientific knowledge, justified the hierarchical relations between the North and the South. Marxist theories, particularly those by Latin American dependency theorists, including Myrdal, Sunkel and Frank, were critical in providing a Southern perspective, showing the utter dependency created between the core and the periphery in order to maintain the status quo for the rich North. Neo-patrimonial structures, an inheritance from the colonial period, further justify state bureaucracy for the purpose of maintaining power relations in favour of the local elite. Under this system, the ruling relationships, both political and administrative, are personal ones. Neo-patrimonialism is characterized by clientelism and corruption. 'Political clientelism is about exchanging or arranging certain services and/or resources in return for political support, such as votes' (Erdmann 2002, 9). Here the focus is on the *arrangement* of services and resources (Erdmann 2002). Popular, local and bottom-up development approaches have brought the debate back to a less fatalistic and more proactive course, where citizens have the power to change.

Under the United Nations' first development decade, the 1960s, development was still measured through the lens of economic growth, with indicators that measure the increase in a nation's gross national product. No attention was paid to the environmental impacts of development or the unequal distribution of the world's wealth. However, the failure of top-down, market-oriented approaches to improve the livelihoods of the world's most impoverished and vulnerable populations impelled the development of bottom-up, community-led development thinking (Boonyabancha 2005; Brohman 1997; Sakai 2002). Decades ago it was obvious that economic development and globalization are insufficient to tackle the problems of poverty, and that the capabilities at the household and individual level needed attention (Sen 1995).

During the 1980s, the Brundtland Report gave rise to a theoretical shift in paradigm with the introduction of the concept of sustainable development, 'the development that meets the needs of the present without compromising the ability of future generations to meet their own needs. Moreover, sustainability is defined as long-term cultural, economic and environmental health and vitality' (City of Seattle 1993, 2). It refers to requirements, demands, and obligations towards future generations and it underlines the responsibility to improve the quality of life of currently deprived and excluded populations.

International political agendas began to recognize the importance of social and environmental needs for successful development. New development indicators, such as the Human Development Index, were created to address these needs – to categorize and map human wellbeing on a global scale. While these new angles and dimensions in development thinking were essential, they were still of a top-down nature and overly quantitative. In 2000 the UN General Assembly prepared the *United Nations Millennium Declaration* with the creation of Millennium Development Goals (MDG), as a global, concerted effort to halving extreme poverty, to halting the spread of HIV/AIDS and providing universal primary education by 2015 (United Nations 2005). International aid and technical co-operation work has increasingly become streamlined, following a global and unified strategy with the development efforts.

The purpose of development is to improve people's lives by expanding their choices, freedom and dignity. New perspectives into the meaning of poverty have, in their analysis, gone beyond financial deprivation to examine social exclusion, voicelessness and wellbeing as well as quantitative economic indicators. Recent approaches are geared towards bottom-up, community-based development; the poor are encouraged to participate in a dialogue that can empower them to create their own definition of poverty, and then create development initiatives suitable to individual and community needs. As theories of development now address environmental sustainability more openly, understanding human development beyond the single aim of economic growth has become imperative. Furthermore, as governments embrace bottom-up participatory development, new foundations are laid out for tackling poverty reduction.

Popular development is '...focused on a central concern of all alternative development approaches – creating development appropriate to the needs and interests of the popular majority in Third World countries' (Brohman 2000, 245). With popular development, communities have gained new interest as a unit for local development initiatives. The terms community, social group, and neighbourhood will be used here as a geographic unit, defined as a community of '..."interest" bound by culture, religion, profession, etc.' (Meyer et al. 2005, 33). The social group or community can also be characterized by the state of deprivation and exclusion, and the struggle for wellbeing and inclusion. The term social group entails '...a collective of persons differentiated from at least one other group by cultural forms, practices, or way of life. Members of a group have a specific affinity with one another because of their similar experience or way of life, which prompts them to associate with one another more than with those not identified with the group, or in a different way' (Young 1990, 43). Furthermore, Young recognizes that groups are an expression of social relations (Young 1990). The discourse recognizes that communities are symbolic and socially constructed (Meyer et al. 2005). The formal and informal social relationships between those living in communities are emphasized as the common denominator applicable to a certain spatial unit. Social and geographic limits of a neighbourhood or a community may be flexible, dynamic and diffuse. The community is the ideal unit for the implementation of concrete sustainable development actions.

Sustainable Development and Livelihoods

Communities and their health can also be analyzed through the sustainable livelihoods approach. This is an assets-based approach to development that focuses on people's strengths rather than their needs. The focus is on empowering people to help themselves, and decreasing the vulnerability of the poor by improving access to and control over livelihood assets (Chambers and Conway 1992; DFID et al. 2002; Mitlin 2003; Rakodi and Lloyd-Jones 2002). While the framework is complex and multifaceted, it is meant to be a flexible tool that can be adapted for the use under numerous development areas. Created as part of a research tool to understand the multifarious dimensions of poverty in rural areas in developing countries, the approach has more recently also been applied to low-income populations in urban areas. This concept goes beyond indicators of household consumption – it considers the perceptions of the people themselves. The approach follows the predominant paradigm with adopting an economic terminology and with highlighting the fact that other assets, such as social cohesion or local knowledge can also be expressed as 'capitals'. The framework integrates the following five livelihood assets, expressed as 'capitals', which are central to the ability of individuals to create and improve their livelihoods:

1. Human capital: labour resources, time, skills, knowledge and health status of an individual or household.

2. Social capital: social networks, relationships of trust, and institutions, which people can rely on for part of their survival.

3. Physical capital: basic infrastructure and equipment, which allow people to engage in livelihood pursuits.

4. Natural capital: the natural resources on which people rely to create their livelihoods, such as land, water, air, vegetation, animals.

5. Financial capital: savings, pensions, income, credit and other financial resources available to individuals to build on their other assets or buffer against shocks.

Political assets permeate the human and social sphere. All these assets influence and are influenced by their level of vulnerability and resilience to external shocks and stresses. Chambers (1989) expanded the understanding of wellbeing and deprivation, including notions of physical weakness, isolation, vulnerability, and powerlessness. Baulch (1996) introduced access to common pool resources, dignity and autonomy as important indicators. Booth et al. (1998) included social subordination, reduced life chances, and excessive workloads as additional forms of deprivation that affect human capital.

The framework is used in the search for adequate strategies on the household or community level addressing their wellbeing. A wide range of mostly qualitative methods and monitoring tools are available to implement the sustainable livelihoods approach, allowing for specific adaptations. A livelihood is environmentally sustainable when it maintains or enhances the local and global assets on which livelihoods depend, and has net beneficial effects on other livelihoods. A livelihood which is socially sustainable can cope with and recover from stress and shocks, and provide for future generations. This framework looks into policies, institutions, and processes, since these definitely can influence or control a person's ability to access their assets and to develop the strategies that can maximize their livelihood outcomes. Governance that enhances the access of the poor to basic livelihoods assets reduces their vulnerability and helps ensure a more equitable use of the resources. The sustainable livelihoods approach does not only focus on their needs; but looks at the existing assets of the individuals or the community and, most importantly, places the people in the centre of the analysis. This approach links the activities that people perform on a household level to sustain their livelihood, to a broader, local, regional and international scale (Bury 2004). This method of integrated analysis can help construct sustainable communities, a continuous process that evaluates, adapts and changes over time.

Poverty: A Continuing Central Problem in the City

Poverty reduction has been a central theme in the international development debate since the late 1990s. Grassroots and NGO initiatives and various UN agencies are engaged in finding solutions to address the structural causes of poverty. The figures on poverty can diverge according to the definition applied for poverty and the information source used. The World Bank shows that the absolute number of extreme poor – defined as people living on less than one US$/day – has declined worldwide between 1981 and 2004 (Chen and Ravallion 2007). Looking at regional trends the data shows that whereas the number of poor has declined in Asia, it has risen in Sub-Saharan Africa and also slightly in Latin America (Chen and Ravallion 2007). When focusing on worldwide inequality, statistics indicate a trend of rising inequality over the past forty years. The Social Panorama of Latin America Report estimates that 36.5 per cent of Latin America's population (195 million people) were still considered poor (less than two US$/day) and 13.4 per cent (71 million) continued in extreme poverty or indigence (less than one US$/day). These numbers signal a 3.3 per cent drop in poverty and a two per cent decrease in extreme poverty, compared to the indicators in 2005 (ECLAC 2007). Brazil registered decreases of 4.2 percentage points in both its poverty and its extreme poverty rates between 2001 and 2006. In 2001 37.5 per cent of the population lived in poverty and 13.2 per cent lived in extreme poverty. In 2005 the numbers were 36.3 and 10.6 and in 2006, 33.3 and 9.0 (ECLAC 2007). This represents a reduction in the number of indigents of 6 million people. The ECLAC study (2007) asserts that the *bolsa familia* public transfer programme implemented in the country has played a decisive role in this achievement.

Definitions of poverty have shifted from a traditional, income-oriented economic approach, as promoted by the World Bank for many years, to one that is multi-faceted and focused on distributive issues of basic needs, like access to adequate health care, education, housing, information or nutrition (Satterthwaite 2001). The analysis now also focuses on qualitative indicators rather than only quantitative ones, a change that parallels the changes in the debate over the concept of development. Recently, poverty analyses use an extended framework that includes dimensions of vulnerability and powerlessness (Hjorth 2003) as described earlier within the sustainable livelihoods concept.

Social exclusion – a complex, socially constructed phenomenon with social, economic and cultural facets – is defined by Room (1995) as a state of detachment, where individuals are restrained from or not enabled to access public services, goods, activities, or resources which are essential for a life in dignity. It is seen as a state of illbeing that '...takes power away and somehow disables people from being full citizens, isolating them from the rest of society' (ILO 1996, 12). Unemployment, underemployment and the consequent impoverishment are the most visible and direct factors that drive social exclusion. Poverty, understood as lack of access to vital economic resources, is the main cause for economic exclusion (Yapa 1998).

The stigmatization of certain groups as being an 'underclass' who are worth less than other groups reinforces social exclusion and provides excuses for not doing enough to achieve greater equality (Gans 1996). Strong prejudice – based on race, education, and income – against people from marginal settlements results in those people being considered as lower in social status. Gender related prejudices often force women into disadvantaged employment patterns and restrictive social roles (Bandarage 1997). Marginalized children and young people are affected most, since they are deprived of future opportunities. Not being able to regularly eat healthy food and access formal education or professional training, perpetuates the cycle of poverty and exclusion. This is true of the thousands of informal recyclers that make a living through separating materials. Social, economic, and political structures – as well as incapacity, omission or corruption – often create and sustain exclusion, ultimately resulting in the denial of citizenship (Fainstein 1996; ILO 1996; Mingione 1996; Room 1995; Walker 1995).

There are also global factors that contribute to social exclusion. The increasing internationalization of production, trade and consumption, particularly since the early 1970s, has made the world economy more complex and more interrelated. Changes in the international division of labour, with increased precariousness of work and labour shifting into regions and countries with the most exploited working conditions, have affected employment around the globe (Lavinas and Nabuco 1995; Bromley 1997). There is growing awareness of the crime and violence implications of joblessness, long-term unemployment, and widespread underemployment. Economic recession and structural adjustment programmes directly impact on public spending, usually encouraging funding cuts particularly in the public health, education, and housing sectors. These consequences further underpin some of the already existing social and economic disparities and disadvantages (Bello 1993, 201–204; Drakakis-Smith 1996, 692). Institutionalized corruption, political clientelism, and short-term measures of populist political nature are also responsible for chronic and widespread social and economic inequalities and growing regional disparities. Continuous socio-economic circumstances of deprivation reinforce the inability of the excluded to overcome this state. 'Material deprivations experienced by the poor are socially constructed at every node of the nexus of production relations' (Yapa 1998, 95).

Sen (1992) focuses on the existing connection between livelihood and citizenship rights. He explains how in different circumstances, individuals become excluded from what are considered basic public goods and services and from basic consumption. Social exclusion is a violation of citizenship rights (for example, it violates the basic right to housing, education or health care). Exclusion increases the potential for crime and violence. It also generates environmental impacts, whose costs are also externalized. We will see in the case study that exclusion is the driving force for illegal occupations and for the consequent environmental degradation, with irregularities in domestic waste disposal, sewage, water and electricity supply. The situation further reinforces environmental exclusion, when natural resources are not protected or adequately managed. The loss of social capital due to crime or violence and the loss of environmental capital due to pollution and degradation are obvious consequences of exclusion. We can therefore presume that exclusion equally jeopardizes social and environmental capital.

The problems related to exclusion are further aggravated by negligence in the educational sector. Lack of good quality education diminishes the prospect of formal employment and the chances of full citizenship. It contributes to social exclusion and ultimately has harmful drawbacks on the environment. Environmental education programmes – aiming at citizenship awareness in the context of an equitable and sustainable society – are powerful tools for social transformation through which exclusion can be diminished. Social inclusion becomes a basic condition for sustainability. Measures to make marginal spaces more liveable need to include environmental awareness actions, from the grassroots level and of the government. The fragmented nature of governance structures with multiple institutional levels is frequently a hindrance for efficient and economic resource management. Lack of communication between agencies, excessive bureaucratic procedures, rivalry, corruption and inefficiencies are among the major institutional barriers (Soto 1989) that prevent more active integrated steps towards inclusion. Widely recognized in the literature, the concept of social exclusion has now entered policy levels, with governments now creating specific agencies and programmes to combat this phenomenon. In good governance, social exclusion is becoming a central theme, finding more sustainable solutions for social urban development issues. It will be discussed in specific contexts throughout several chapters of this book.

Territorial Justice

The feminist model sees justice as according respect and participation in decision-making to those who are dependent as well as to those who are independent. '*Social justice*' more specifically concerns the degree to which a society contains and supports the institutional conditions necessary for the realization of these values: 'developing and exercising one's capacities and expressing one's experience and participating in determining one's action and the conditions of one's action' (Young 1994, 37). The distributive paradigm defines social justice as the morally proper distribution of social benefits and burdens among society's members. Social justice is about distribution of wealth, income, material resources, rights, opportunity, power and self-respect.

Since the 1960s the concept of justice has been applied to the analysis of many urban spatial patterns, by focusing on variations in the geographic distribution of goods and services (Kirby and Pinch 1983). In other words, distributive justice is about how fairly the cake is divided among its members (CSJ 1998). In the past, a utilitarian approach to measuring the distributional outcomes gave rise to the notion of *territorial justice*, as a measure of fairness in the distribution of accessible public services (Davies 1968) and related to the proximity to undesirable land uses (Dicken and Lloyd 1981). Important debates on fairness with respect to the distribution of environmental quality and risks emerged in the USA during the 1980s. Low and Gleeson point out that '...the distributions, which are highly variegated in socio-cultural and spatial terms, interact to produce a diverse and shifting landscape of ecological politics' (2001, 104).

Social exclusion becomes apparent in the widespread informal activities and the spread of informal housing. In his book *The Other Path*, Hernando de Soto (1989) shows that, in Peru, informal acquisition of land property by the poor is the only way to guarantee their livelihoods. Here informality is defined as '...refuge of individuals who find that the cost of abiding by existing laws in the pursuit of legitimate economic objectives exceeds the benefits' (de Soto 1989, xxi). The analysis further shows that land invasions and occupations follow organization; they are not chaotic, though they are often perceived as such by the general public through a prejudiced view. Some of de Soto's ideas on poverty and injustice in Peru also explain the situation in other developing countries.

The term squatter settlement refers to '...uncontrolled, low-income residential areas with ambiguous legal status regarding land occupation...[and the] usual image of a squatter settlement is of a poor, underserviced, overcrowded and dilapidated settlement' (HABITAT 1982, 15). These are spaces where access to drinking water, sewage, waste collection, and other basic public services are mostly improvised; where spaces for leisure and cultural activities are rare or non-existent and where housing is risk prone. If not lacking, the services and infrastructure provided are often in such precarious condition in these neighbourhoods that public and environmental health are at stake. Unplanned and uncontrolled squatting causes social and economic impacts. Open cut sewage and irregular dumping of garbage affect the quality of the drinking water in nearby reservoirs. Due to their illegality, the population of these settlements has to compete for the public services with neighbouring populations. Water shortage and conflict over water are increasingly the consequences.

Because of their excluded condition and the absence of basic infrastructure, the dwellers generate environmental health impacts, which then directly impact on themselves. The dwellers are often defenceless because they lack resources, information, and formal education. Although they also have a share of co-responsibility they are constantly exposed to risks and hazards in their living environment. The correlation between location of low-income neighbourhoods and environmental hazards, such as disease vectors (rats, insects), waste dumps, polluting industries, and contaminated streams and lakes, provide proof of environmental injustice. Urban peripheries are mostly treated as excluded spaces, where environmental legislation and regulations to control land use, deforestation, and the discharge of contaminants are less enforced or are ignored. Hence, the sustainability debate needs to equally address social and environmental exclusion.

Housing Crisis at the Urban Fringe in São Paulo

One of the most populous regions in the world is the Greater Metropolitan Area of São Paulo with approximately 20.5 million inhabitants spread in 39 municipalities that continuously cover an area of 8,051 km² (IBGE 2005). The city of São Paulo is the largest within this agglomeration. It is well known for its skyrocketing population growth – from 580,000 inhabitants in the 1920s, to almost 1 million in 1930 and to 10 million in 2000. In 2005 there were 10.9 million people living in the city occupying an area of 1,523 km² (IBGE 2005).

Socio-economic inequality has always been a reality in São Paulo. Historically, the mansions of rich coffee barons and landowners along what is now called Avenida Paulista, the most expensive locations of the city, were surrounded by poor slaves and servants. Today powerful entrepreneurs and global investors live side-by-side with *favela* inhabitants. The city displays an acute disparity in the income distribution of its population. The gap between the richest and the poorest is increasing and so are the absolute numbers of the poor. It is a city of extremes; it hosts wealth, cultural diversity and technology and it has well-skilled labour to offer. The phenomena is part of the contemporary drive for a global market orientation; which generates inequality with widespread areas of poverty, particularly in the urban periphery. Increasing unemployment, informal jobs, environmental degradation, crime, violence and abandonment become visible in these spaces.

Although, the overall annual population growth rate in São Paulo has diminished from 4 per cent in 1991 to 1.7 per cent, the city is still growing very fast at the fringe with over 6 per cent annual increase since the early 1990s. Over 30 per cent of the city's population lives now in marginal settlements, classified as peri-urban (Torres et al. 2005, 1). These areas are characterized by continuous high population increase (Torres et al. 2005, 16) as well as by social and environmental exclusion. Despite the wealth generated in the metropolitan region of São Paulo (16.7 per cent or US$ 99.1 billion of Brazil's total GDP in 2000), there is a sharp contrast between the spaces for the better off and the spaces for the excluded.

Caldeira (2000) explains that the first driving force in generating a new pattern of urbanization at the end of the 1930s was the expansion of the urban bus transport system. Private entrepreneurs – most of whom were also real estate speculators – rather than the government determined the early territorial expansion of the urban fringe. There were no official development plans until the 1970s, when the periphery became transformed from rural into largely unregulated urban spaces, open to exploration and exploitation by the private initiative. It was a situation comparable to the *Far West* with unregulated speculation, fraudulent practices, profit maximization and at the same time without enforcement of environmental protection (Caldeira 2000).

Without adequate financing systems that would allow the poor to acquire homes, irregular occupations became the only option for many of the poor in the city. The requirements of the few existing lending programmes for social housing were impossible to fulfil by the poor. As a result, it was the middle class that used these programmes, as happened with the BNH programme (Banco Nacional de Habitação) (Caldeira 2000, 224). The desperate need for housing, as a consequence of the population increase, and the lack of financial capital, gave rise to the homeless movement (Movimento dos Sem Teto) in the early 1970s. This movement was important in influencing the government to invest in infrastructure programmes, particularly sanitation in the periphery. More recently, since the 2000s, squatting and the occupation of abandoned buildings have become organized practices by local and national homeless movements (Movimento Nacional dos Sem Teto, União Nacional por Moradia Popular (UNMP), Movimento dos Trabalhadores sem Teto (MTST), União da Luta pelos Cortiços, and others). Today the national movement UNMP unites more than 150 smaller movements throughout Brazil. Given the widespread housing deficit, most large cities have a homeless movement. According to figures mentioned by the movement, there is currently a deficit of 7 million homes in the country. The movement defines homeless as those people who have no dwelling, who live in substandard and overcrowded housing circumstances, or share the home with family or friends as a favour. Thus, it unites those people that do not have sufficient money to pay for rental: all those who are without a dwelling as a consequence of their precarious social condition. Census data show that in 2000 there were 16.5 million people in Brazil depending on favours or irregular squatting (Miranda 2007). Under the current economic circumstances the poor are unable to pay the cost of formally accessing land, infrastructure and housing.

This new development at São Paulo's fringe since the early 1970s has led to widespread, irregular squatter settlements with self-constructed houses, precarious or no basic infrastructure, and the absence of environmental protection. The term *squatter settlement* (in Brazil known as *favelas*) refers to the dwellings where the poor have occupied land – sometimes paying for it – without legal ownership or planning permission (Alsayyad 1993). The origin of the term *favela* stems from the denomination of the first occupation in Rio de Janeiro in 1897, called Morro da Providência, later called Morro da Favela, after the demolition of the local *cortiço*, a typically dense poor housing areas in the city centre (Valladares 2005). *Favela* stands for '…conjunto de barracos aglomerados sem traçado de ruas nem acesso aos serviços públicos, sobre terrenos públicos ou privados invadidos' [a group of squatters build closely together, without planned streets or access of public infrastructure, occupying public or private land] (Valladares 2005, 26). The National Bureau of Statistics (IBGE) refers to *favelas* as illegal agglomerations of more than 50 *substandard homes*, built without permission and with improvised construction materials; in the international literature known as *squatter settlements*.

It was only during the mid-twentieth century that the *favela* phenomena began to spread in most large cities. By then the term was widely accepted by the mainstream, and was used to categorize a poor habitat form, deriving from illegal and irregular occupation, without respecting norms or planning regulations (Valladares 2005). These settlements usually spread on lower priced land, often on steep hillsides, or in areas prone to flooding. Valladares also shows how the term has entered the imagination of the public, interpreting *favela* as a stigmatized place, an anti-hygienic habitation form infested by *vagabundos e criminosos* [vagabonds and delinquents] (Valladares 2005, 26 and 39). *Favelas* were seen as problems and its inhabitants the *favelados* as troublesome. They became the spaces for the socially and economically excluded.

The number of people living in unregulated housing conditions in São Paulo has increased from 71,840 in 1973 (1.1 per cent of the total population) to over one million in 1991 (11.3 per cent of the total population) (Taschner 1995, 193). One result of this widely unregulated urban development is the relatively high rate of home ownership in the periphery of São Paulo: 68.5 per cent compared to the city's average of 63.7 per cent (Caldeira 2000, 224). Not all of the irregular housing can be considered *favela*. The irregularity of these occupations can bear serious health consequences, because they often lack basic infrastructure (Potter and Lloyd-Evans, 139; Taschner 1995). Until recently urban planning and infrastructural improvements were limited to the central areas of the city. Fix (1996) has analysed government-funded, luxurious urban developments in privileged areas in contrast to the infrastructural deprivation witnessed in the periphery. The results show a general increase in social and territorial inequalities for the city of São Paulo.

In the 1990s housing arrangements became further fragmented and disparate, as a consequence of economic recession, unemployment, growing living expenditures, and the demand for cheap housing during the 1980s and early 1990s. In 2000, approximately 10.2 per cent of the population in São Paulo was living under precarious conditions, with 870,000 people living in *favelas*, 182,000 living in *cortiços*, and 8,704 homeless living in the streets and below bridges (IBGE 2005). The most precarious and inhumane housing alternative for the poor are so-called *cortiços*. The *cortiço* is defined as permanent improvisation, with more than one family sharing one apartment or home under precarious sanitation and very dense living conditions. Valladares (2005) describes them as spaces of urban poverty that have existed much before the *favelas*. *Cortiços* are found in the city centre but also in the poor periphery, where dwellers sometimes add another room for renting on top, as an additional income (Kowarick 1988).

Lack of affordable shelter causes homelessness for many of the poor. Many homeless families have children, which is one of the explanations of children living in the streets of São Paulo. At the beginning of the twenty-first century, 1.25 per cent of children below age six and 4.6 per cent between ages seven and 17 are homeless (Prefeitura do Município de São Paulo 2000, 61). The precarious living conditions increase the risk that these children become involved in drug trafficking, drug consumption and crime, resulting in unnecessary loss of life.

Far from being homogeneous spaces, *favelas* are diverse in their physical, spatial and social characterization. A study in marginal areas in Rio de Janeiro reveals that there is great diversity related to: a) construction and territorial characters; b) urban infrastructure (water, sewage, waste collection); and c) formal education and income of the household heads. Most *favela* dwellers live in misery. This is a characteristic of the population in other sub-standard urban housing conditions, such as *loteamento periféricos* [low-income new housing area in the periphery] or *bairros populares* [popular neighbourhoods] (Valladares 2005, 157).

Environmental Conflicts: Occupations in Protected Watersheds

Irregular occupations in the protected watersheds of Lake Billings and Lake Guarapiranga in the south of São Paulo, are particularly alarming (see Figure 3.1).

The total population living in the Billings watershed for example has increased from 116,000 to 495,000 between 1970 and 1996 (Estado de São Paulo 1997). Today there are approximately 700,000 people living in the Billings watershed, of which over 121,000 live in *favelas* (Capobianco and Whately 2002, 17).

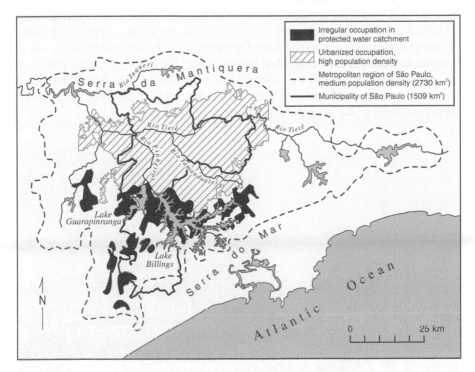

Figure 3.1   Urban expansion into protected water catchment

Sources: Capobianco, J.P.R. and Whateley, M. (2002), *Billings 2000* (São Paulo: Instituto Socioambiental).

Cartography: Ole J. Heggen.

For decades, the Billings catchment had received sewage and wastewater from the rivers *Tietê* and *Pinheiros* and their tributaries. Garbage is being washed into the lake from the drainage and streams of the surrounding neighbourhoods. The specific environmental protection legislation for drinking water basins, from 1975/76 (state law no. 898/75 and 1.172/76), was never strictly implemented. Only since the introduction of the state law 9,866/97, which provides specific instruments and mechanisms for decentralized land use management and protection, have interventions for environmental rehabilitation in the area been allowed. This legislation also permits greater participation of the local community and grassroots through public hearings, partnership projects and other instruments included in the *City Act* and the city's *Master Plan* (Alfonsin 1997, 199–219; Estado de São Paulo 1998, 13).

Insufficient alternative housing programmes and social policies, in addition to a lack of political will and the absence of environmental law enforcement under past local governments, have been major driving forces for irregular urban sprawl, including sprawl in environmentally protected areas (see Illustration 3.1). The fragmentation of public agencies and policies as well as the lack of co-ordination and co-operation among them is another key reason that has led to the absence of preventive and mitigating strategies. Squatting was tolerated and sometimes even promoted as a measure to decrease the pressure on the social housing sector. Even today dwellers are relocated from prime locations in the centre of the city to the periphery (Fix 1996). For the expansion of a major road Água Espraiada, a local *favela* was removed and its inhabitants were resettled at the Mata Virgem watershed in the periphery, one of the last areas which was covered with native forest in the south of São Paulo. The settlers received a small compensation, but were left without technical advice or infrastructure support to dwell on the steep and forested slopes (Fix 1996).

Illustration 3.1   Irregular urban sprawl at Lake Billings

One alternative for the growing proportion of no- and low-income population is to occupy land. Sometimes these invasions (*invasões*) are co-ordinated by sect leaders, for example the Evangelical Pentecostal church (*Assembleia de Deus*); by local politicians (often the district councillors); or by landowners who expect the city to improve urban services in the area. In exchange for their votes or support, the leaders or individuals protect the settlers from eventual eviction. Real estate agents frequently also claim and sell land illegally. The following section will examine some the particular challenges that persist in marginal settlements, taking the example of the *Pedra sobre Pedra* community.

Case Study: Liveability on Hold in the Periphery of São Paulo

Squatting in the watersheds south of São Paulo, first began near the margins of Lake Billings during the late 1970s, on public land owned by the then state electricity company (*Eletropaulo*). There were also several quarries in the area for sand and stone extraction. In 1980 extraction from the Itatinga quarry stopped and the area was then used as a landfill for inert materials. At the same time, part of the land gradually became occupied; five years later, around 100 families were already living next to this new landfill. In the early 1990s squatting continued along the unconsolidated steep slopes and the lower areas of the quarry. The plots were small, approximately 20 to 50m² per family. As in many other locations the municipality was unable to control land invasions and occupations of public land, along the margins of the creeks and the lake. The dwellings rapidly expanded during the 1990s, from the abrupt hillside to the limits of the dumping ground (see Illustration 3.2).

The neighbourhood association *Associação Pedra sobre Pedra* (APSP) was created in 1991 to assist the dwellers in their demands for basic infrastructure improvements. According to the leader of APSP, usually more than 100 people from the local community participated in general assemblies that were occasionally organized to address community issues. Few of the people were actually committed in continuous collective action, an involvement that with time further diminished. By the mid 1990s two neighbourhoods *Pedra sobre Pedra* and Maça do Amor were

Illustration 3.2
Itatinga landfill and *Pedra sobre Pedra* squatter settlement

consolidated in the area (Figure 3.2). According to APSP from the people living in these two settlements approximately 4,000 were unemployed or underemployed, 650 children were out of school and 120 people had physical disabilities.

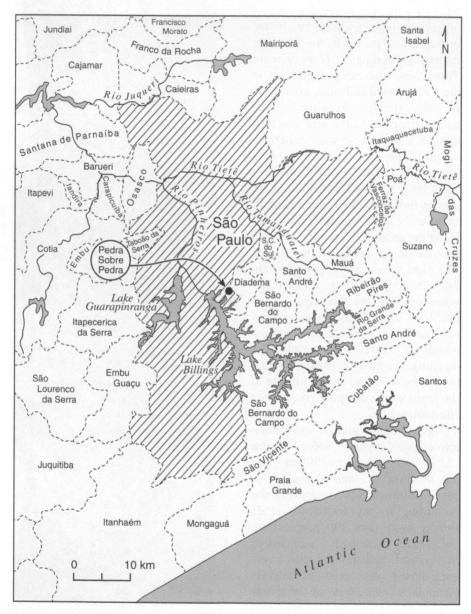

Figure 3.2   Localization map highlighting *Pedra sobre Pedra*

Sources: CPLA/SEMA 1997 and IBGE 1991.

Cartography: Ole J. Heggen.

In 1995/96 APSP conducted interviews with 361 randomly selected heads of households in order to assess local livelihood characteristics. In 1999, when I was working at the Department of Geography at The University of Newcastle, Australia I received funding from the Australian Research Council to carry out a rapid urban appraisal in *Pedra sobre Pedra* settlement, together with APSP leader and Ruth Takahashi, a social assistant who had worked for several years in the area. Our objective was to map risks, hazards, basic infrastructure and public services in the neighbourhood. Seven volunteers from APSP and from another community association (Sociedade Amigo do Balneario Mar Paulista) also collaborated in the appraisal. The leader accompanied the fieldwork from the beginning to the end. We used the following data collection techniques: randomly sampled semi-structured interviews with dwellers, group discussions with members from the local neighbourhood association, structured interviews with municipal government officials from the environmental secretary (Secretaria do Meio Ambiente) and the agency for sanitation and basic infrastructure (SABESP). We conducted several preparatory meetings with APSP and the other volunteers who wanted to participate in the survey. We then explained the objectives and methods and together defined the indicators for quality of life for the survey. We also held a general assembly in the neighbourhood at the beginning of our work, to explain the research objectives and to refine the methodology. Another assembly was conducted at the end of the fieldwork to provide first results and gather local feedback.

In addition to our own data, a previous household survey conducted by members of APSP also provided interesting information on the livelihoods of the inhabitants. In this survey 361 households were asked about formal education, income, employment, housing density, and perceived risks in the neighbourhood among other indicators. The results show the widespread lack of financial capital in the community. When asked about their socio economic situation, 54 (15 per cent) of the 361 households reported financial hardship and 25 (6.9 per cent) mentioned not having enough food at home. In 24 cases (6.7 per cent) the head of the household was unemployed and in 13 cases (3.6 per cent) they were illiterate. Each household had on average 4.4 dependents and 2.2 children. In 63 per cent of the respondents the head of the household was a woman. Figure 3.3 illustrate the precariousness of the financial capital and Figure 3.4 highlights the vulnerable human capital in *Pedra sobre Pedra*.

As part of the rapid appraisal, I also conducted a mapping activity with seven volunteers from APSP and another neighbouring community association, Sociedade Amigo do Balneario Mar Paulista. For three days we walked through all alleys and pathways experiencing and mapping the neighbourhood. We talked to residents, took photographs, mapped the infrastructure and observed the risks and hazards. The research identified areas under risk of landslides and flooding, localized environmental impact – such as irregular garbage deposits, sewage discharge, storm water emission, leaking fluids from the waste dump – as well as areas lacking basic infrastructure, and public facilities. Amongst other indicators, we reported on housing conditions. Most houses were built with bricks (1,276 houses), followed by a significant number of houses made with mixed or improvised building materials (1,038 dwellings). 56 houses were still under construction and 114 houses were already up for sale. In Rua Javali, for example one house was on sale for Reais $5,000 (at the time US$2,525). Brick houses are more resistant and less likely to

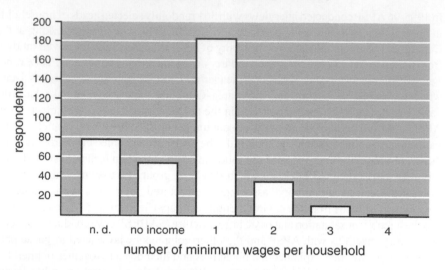

Figure 3.3   Income variation in *Pedra sobre Pedra*

Source: Grupo Técnico de Apoio (1999), *Projeto PROSANEAR: Pedra sobre Pedra e Maçã do Amor*. São Paulo (report).

Cartography: Ole J. Heggen and Jutta Gutberlet.

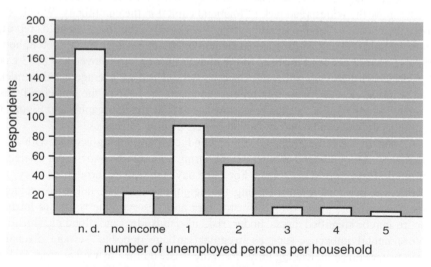

Figure 3.4   Number of unemployed household members

Source: Grupo Técnico de Apoio (1999), *Projeto PROSANEAR: Pedra sobre Pedra e Maçã do Amor*. São Paulo (report).

Cartography: Ole J. Heggen and Jutta Gutberlet.

be torn down by the Government. There were only six vacant plots in the entire neighbourhood and very little green spaces. Almost every centimetre of land is claimed and used. We counted 64 grocery and hardware stores and 26 pubs, followed by hairdressers, electrician services, mechanic services, recycling depots, hairdressers, car repairs and informal ticket sales. Bus tickets were sold to the North and Northeast only, the part of the country where many of the dwellers were originally from. There were also signs offering services in legal assistance, tailoring and homemade foodstuff. We counted 11 churches and visible temples. With every new settlement new opportunities for small-scale business and handicraft work arise. Finally, these settlements also provide a niche for drug dealing and other illegal activities as was pointed out recently in an interview with two leaders from *Pedra sobre Pedra*. Unemployed young people often make a living by dealing drugs. 'It is the drug that commands our neighbourhood, but it does not hinder anybody from inside the community…Today the situation has improved and people don't kill each other any more because of drugs. Today it is calm, you can come and go without any problem' (Interview with Aninha, 13 August 2007). Aninha who lives in *Pedra sobre Pedra* and is the leader of the recycling group Selva de Pedra further describes unemployment, drugs and the corrupt police as major problems in her community. According to Jocemar, the main issues are high unemployment and drug trafficking (Interview with Jocemar, 13 August 2007).

While we were mapping the community, heavy summer rainfalls destabilized one of the slopes and a landslide took several houses, killing one child. The lower situated dwellings were inundated for several hours with rainwater and sewage, displacing families and damaging their belongings. These events always bear the risk of spreading infectious diseases, particularly *Leptospirosis*. Similar hazards occur regularly in many of the precarious settlements in the periphery of the metropolitan area of São Paulo, particularly during the periods of heavy rainfalls in the summer. Due to the instability of the infrastructure and because of governmental omission environmental capital has been seriously damaged, posing unnecessary hazards for the inhabitants. For example, a freshwater spring in this watershed was paved by squatting and the natural vegetation coverage that had contributed to slope stabilization was destroyed, creating environmental hazards.

Living conditions in the periphery are very crowded and there is no space dedicated to recreational purposes. The absence of vegetation becomes a frequent characteristic in these settlements. Occasionally some dwellers grow vegetables, herbs or medicinal plants on the limited space. We counted only ten unoccupied plots and only ten full size trees in the entire area. There is literally no free space in the settlement and expansion happens vertically. Many of the houses have built a second or third floor on top and sometimes houses are subdivided into several units, with several families sharing the space.

Due to its illegal and unregulated status, at the time of the research there was no official supply of basic infrastructure and public services within the settlement. There were no public schools, pre-schools, health care centres, or supermarkets. The informal entrance to the settlement becomes inaccessible during the summer rainfalls. The population relies on the adjacent formal neighbourhood to access water, electricity and garbage collection. Improvisation contributes to water leakage.

There are still a few wells in use, but the risk of contamination is high. As in most other marginalized settlements in São Paulo there is no proper sewage collection; grey water is diverted into the creeks, drainage channels, roads and alleyways, often causing soil erosion.

Electricity is usually 'borrowed' from the serviced neighbourhood. The services in the surrounding working class residential area are already overcrowded. It is very common to see hundreds of cables connecting to the few poles along the main road, from where power is then distributed throughout the settlement. Power cuts are frequent because of short circuits, and the lack of safety causes risks to many people.

Squatter settlements are considered illegal and therefore solid waste is not collected. The population deposits the garbage at the entrances of the settlement, at unoccupied land, or into the drainages. Dwellers frequently complained that others would just throw their garbage downhill, often on top of the roofs or between the houses. 'People used to throw the garbage bags out of their windows. Sometimes the plastic bags become trapped in the vegetation or were carried downhill with the flow of the rainwater' as one of the dwellers puts it (interview conducted in 2000).

Dumping has been identified by many interviewees as causing slope instability and health threats. In fact, the massive accumulation of garbage as well as the open discharge of used water and sewage has contributed to the formation of landslides and floods in the area. Waste also attracts and maintains rats and insects; however, removing the waste without dealing with the population of these animals means they are likely to attack humans, an observation voiced by one of the government agents (Interview with Environmental Secretary, 20 January 1999).

All 22 roads and 112 alleyways in this settlement were unpaved and had no public lighting at the time of the survey. The three main roads were deeply eroded and could only be accessed during dry weather. Narrow and winding alleyways throughout the agglomeration link most dwellings with the main passages. The existing, already insufficient road and public transport system in the surrounding areas collapses during rush hours due to the increase in population and vehicles.

The state of permanent social exclusion (from childhood, from education, from employment with fair pay, from information and knowledge) accounts for the generally low environmental awareness among the population. More pressing issues are: underemployment, unemployment and illegal tenure and the risk of being evicted. The daily struggle is how to survive and how to guarantee the subsistence for the rest of the family. When asked about major livelihood threats 24 interviewees mentioned unemployment, 49 financial difficulties, 23 lack of food, three had all kinds of difficulties and only three answered not having livelihood concerns. The data shows an extremely vulnerable population, on the edge of survival. Their resources are limited reflecting in low human capital.

Social exclusion affects the social capital of the local population, in that the people lack the power to lobby the government for infrastructure improvement. Instability also in terms of housing permanence is another factor that undermines the social capital of this population. Only few of the local residents participate in collective actions with the potential to create infrastructure improvements. The local neighbourhood association (APSP), for example, is struggling to maintain and expand its participation rates.

Once the survey in *Pedra sobre Pedra* was completed and first results were available, a general assembly was held with the presence of local authorities, participants from local NGOs and APSP as well as the public. The event was an opportunity to talk about environmental issues and to communicate about the local recycling initiative. We displayed piles of compressed plastic bottles and aluminium cans and emphasized the importance of public participation in the process of recovering resources and cleaning up the environment. Finally, an open debate on environmental education and social exclusion was held, taking these social and environmental themes into the wider public debate. For the community leaders, this was the beginning of a long struggle for social justice focused on improving the livelihood conditions in this settlement.

Instruments and Policies for Proper Urban Development

Local action needs to be linked to a set of public policies and formal tools that enable the implementation of neighbourhood improvements and local development initiatives. Until recently, public participation in planning and in implementing actions to improve the liveability of the city space were non-existent. However, a first milestone addressing urban planning was set with the incorporation of a chapter on urban policy in the Federal Constitution of 1988 (Fernandes and Rolnik 1998, 41), which requires cities with more than 20,000 inhabitants to elaborate a master plan that considers the potential effects of urban expansion and development. In article 183, it approves the right to '...possession in private urban landholdings up to a maximum of 250 m² after only five years of peaceful, uninterrupted possession of the property, a condition termed *Usocapião*' (Fernandes and Rolnik 1998, 147; see also Brazil 2001, 167). Although this measure was created to slow down urban mobility by providing for a more stable land tenure situation, without law enforcement and without providing housing alternatives, it has rather resulted in the increase in illegal occupations and a drive to land speculation at the fringe.

Important changes in urban development are underway with the City Act (Estatuto da Cidade) a federal law enacted by the National Congress (Law No. 10,257/2001) as well as Municipal Law No. 15,547/1991 (Brazil 2001; Bassul 2002). This act proposes participatory decision-making strategies in urban development (Article 20.II). The City Act provides the opportunity for legitimized participation of the community through neighbourhood or community associations and environmental NGOs (Article 50). It also stimulates citizens to initiate public action through the Public Ministry when collective rights are infringed. Neighbourhood associations with a mandate for environmental protection can thus take an active role in urban planning, by way of '...the right to participate in decision-making allows the creation of institutional channels that can serve as spaces to mediate and negotiate conflicts' (Caldeira 2000, 147). These legal spaces are important also to disseminate knowledge about citizen rights. Justice requires participation in the processes of democratic decision-making. As to Young (1994), '...all persons should have the right and opportunity to participate in the deliberation and decision-making of the institutions to which their actions contribute or which directly affect their actions' (91).

There are also procedural juridical instruments that assist the state in intervention against violations to the principles of urban development. The City Act further allows for legal usage of taxation or even expropriation in order to fulfil social purposes (Meneguello 2002; Brazil 2001). Land tenure contributes to the recognition of citizenship and fosters a sense of place among the dwellers. Consequently, it has the potential to enhance the environmental stewardship in the community.

In 2002, the city council also enacted the long awaited Master Plan (*Plano Diretor*) for São Paulo. The Plan, largely supported by the City Act, has to '…fulfil the constitutional premise to guarantee the social function of the city and of urban property' (Brazil 2001, 43). It redefines urban land use zoning and focuses on the following major principles: (1) Act in solidarity towards excluded populations. (2) Consider housing as a social right. (3) Complete and expand roads and transportation systems in the periphery. (4) Salvage the urban environment. (5) Transfer funds from developers to public works. (6) Strengthen the public sector's initiatives and planning projects (Brazil 2001, 40). These two pieces of legislation provide new opportunities to correct the disparities in urban development existing between core and peripheral areas and to implement actions to enhance urban liveability and sustainability throughout all parts of the city.

There have been several attempts to upgrade some of the squatter settlements. In the late 1990s, the *Guarapiranga Project*, funded by the Inter-American Development Bank, tackled urban development and environmental education. The project has achieved some visible improvements in housing, risk minimization, and basic infrastructure. Another project, *ProSanear*, funded by the Federal Government and executed by the state water agency SABESP, has addressed some of the environmental problems in the Billings watershed, including the neighbourhood *Pedra sobre Pedra*, in the early 2000s. Under this project the drinking water supply has been expanded into some squatter settlements and in 2002, under *Projeto Tietê* (IDB-SABESP funding), sewage collection was introduced in some parts of the urban periphery in the Billings watersheds.

Conclusion: Living in the Urban Periphery

The case study provides insights on livelihood conditions and human security threats at the periphery of São Paulo. It portrays the situation lived in many cities in Brazil and in many other countries in the South. It discusses social exclusion, poverty, land speculation, local politics and lack of social housing as the driving forces for the expansion of irregular squatter settlements. One characteristic of these settlements is the lack of basic infrastructure. Besides impacting on the dwellers' quality of life, this causes environmental health problems. Squatters are confined to unhealthy and crowded living conditions. Disease vectors multiply rapidly under these circumstances, increasing the risks of disease transmission and epidemics. The risk further increases when the existing health care system is not able to rapidly and effectively respond to possible outbreaks.

Infectious diseases continue to be the world's leading causes of death, and the danger increases as new diseases continue to emerge. Since 1973, for example, 29 previously unknown diseases have been identified. These can become significant public health threats. There is also the risk that previously controlled infectious diseases, such as cholera, dengue or yellow fever, will emerge again (WHO 1999, 146).

Public policies, especially those dealing with environmental protection, have widely bypassed urban development in the periphery, leading to an unprecedented social and environmental crisis, as we have witnessed in the Billings watershed. Poverty, unemployment and social exclusion are widespread in these settlements, often stimulating crime and illicit drug consumption and trafficking. Environmental degradation as consequence of the urban expansion is obvious in this watershed, with deforestation, erosion, discharge of untreated sewage and solid waste, air pollution from garbage incineration, and intensification of the local traffic. There are no green public spaces in this settlement, further disconnecting people from nature. The public housing, health and educational sectors need to re-define priorities, pinpointing deficiencies and assets for change. Even more importantly, opportunities need to be created for the locals to participate in the shaping of the policies that can influence them.

Peri-urban spaces are growing continuously and rapidly because of the high intra-urban mobility and immigration from poorer regions in Brazil, particularly from the Northeast. Public policies are not always efficient in tackling the urgent emerging social and environmental issues. Unresolved land tenure makes it difficult for squatters to create a sense of place and to build social cohesion, which affects the social capital of the place. Furthermore, because individuals have not had the chance to participate in formal education and are not able to access information, the level of environmental education is low. Yet programmes that build understanding for environmental issues are essential tools for social transformation and need to be integrated in local development.

For years, the leaders of the neighbourhood association in *Pedra sobre Pedra*, were engaged in the struggle to obtain recognition and support to improve the quality of life in their community. Persistent pressure on the local government, and empowerment through research and collective actions can help the community draw the government's attention to addressing the precarious living conditions. Communities need to be able to access funds and information on appropriate urban development. It is difficult for a neighbourhood association to implement self-help measures. Finding the necessary financial resources requires time and resilience, and yet often communities operate on the edge of their livelihoods and do not have the assets to engage in these struggles.

In 2001, the local government implemented a programme to regulate access to electricity and drinking water in the area (*Projeto ProSanear*). Water is becoming a scarce resource in São Paulo and has to be rationed in many neighbourhoods, particularly during the dry winter months. Inefficiency is one of the reasons for the waste of this resource. In São Paulo, 42 per cent of the drinking water is lost mainly due to inefficient and unaccounted water supply (SABESP 1997). Water leakage is common and often a result of the lack of maintenance and irregular connections to piped water. The implementation of this government programme has promoted regular connection to drinking water and energy supply. Several access stairs and drainage channels for storm water along the main pathways have been built since.

Some of the households sited in areas under risk were relocated. However, as of today early 2008, neither the park, the plant nursery, the community garden, nor the centre for environmental education proposed by the neighbourhood association have been implemented.

Participatory and bottom-up development strategies redefine who decides and who applies urban development for whom (Mazzucchelli 1995; Amin and Thrift 2002). It is the responsibility of local governments to involve communities in the assessment and implementation of measures to fulfil their infrastructure and social services needs. It is also an obligation of the global community to respond to deprivation with actions that redistribute resources. It requires international pressure to change the structures and procedures responsible for the current situation of uneven development. The following Chapter 4 will show examples highlighting possible strategies to address social and environmental exclusion.

It is well documented that bottom-up, participatory approaches are more complex and take longer in their implementation. According to Zanen and De Groot (1991) the enhancement of participation first of all requires that people be treated as knowledgeable and trustworthy actors, valuing local knowledge. It is usually more likely to hear the voices of the most engaged individuals, which typically are also the most political and most well educated actors (Carley 2001). Grassroots approaches demand ample ongoing negotiations between the different stakeholders, each of whom may have different expectations and necessities. This process is built on partnership. Often the research process itself can impact on the stakeholders and community members involved, as it raises problems, identifies assets and discusses appropriate solutions with them. Action-oriented research is based on local participation, empowering the stakeholders and strengthening local autonomy (Mahon et al. 2003). Young highlights the intrinsic value of 'participatory democratic processes' as being '...the best way for citizens to ensure that their own needs and interests will be voiced and will not be dominated by other interests' (Young 1994, 92).

As Freire notes '...it is in the public arena that a great variety of people can meet, the habit of association can develop, and the roots of democracy can be cultivated' (Cited in: Saraví 2004, 34). Public space and social practices generated at the neighbourhood level can provide the basis for collective action. The segregation between poor and rich spaces hides the extent and level of poverty. Not everybody is exposed to the visible poverty in the urban fringe, and often the elite is confined to the spaces of wealth, behind protective walls. In São Paulo, peripheral areas, with extremely high rates of population growth, precarious living conditions, and high levels of environmental degradation are usually not on the top of the political agenda. Political and administrative decentralization based on smaller administrative units (*sub-prefeituras*), provide better conditions for local development approaches, such as participatory budgeting or community representation. Public hearings, general assemblies, participatory councils, and inclusive budgeting have a greater potential for creating sustainable livelihoods. Finally, urban movements, neighbourhood associations and co-operatives are important vehicles empowering citizens and challenging conventional development approaches in favour of more participatory planning and decision-making that improves overall sustainability.

A paradigm shift is necessary to attain long-term growth in quality. Development needs to be re-thought as processes towards the construction of sustainable communities with bottom-up approaches, involving the local stakeholders. 'Democratic procedures and governance in general, as well as in urban settings already do rely and will continue to rely...upon the mediating institutions of local action and the formation of local solidarities' (Harvey 2001, 207). Interactive learning as well as participatory community-based approaches are able to provoke effective change towards increased sustainability and liveability on a long-term basis.

References

Alfonsin, B. de M. (1997), 'Direito à moradia: instrumentos e experiências de regularização fundiária nas cidades brasileiras', *Observatório de Políticas Urbanas*, Rio de Janeiro: IPPUR-FASE (Report).

Allen, J., Massey, D. and Pryke, M. (1999), *Unsettling Cities Movement Settlements* (London: Routledge).

Alsayyad, N. (1993), 'Squatting, culture, and development. A comparative analysis of informal settlements in Latin America and the Middle East', *Habitat International* 17(1), 33–44.

Amin, A. and Thrift, N. (2002), *Cities: Re-imagining the Urban* (Cambridge: Polity Press).

Arce, A. (2003), 'Value contestations in development interventions: Community development and sustainable livelihoods approaches', *Community Development Journal* 38(3), 199–212.

Bandarage, A. (1997), *Women, Population and Global Crisis. A Political-economic Analysis* (London: Zed Books).

Barkin, D. (1997), 'Will higher productivity improve living standards?' in Burgess, R., Carmona, M. and Kolstee, T. (eds), *The Challenge of Sustainable Cities: Neoliberalism and Urban Strategies in Developing Countries* (New Jersey: Zed Books).

Bassul, J.R. (2002), 'Reforma urbana e Estatuto da Cidade', *Eure* 28(84), 133–44.

Baulch, R. (1996), 'Neglected trade-offs in poverty measurement', *IDS Bulletin* 27, 36–43.

Baumann, P. and Sinha, S. (2001), 'Linking development with democratic processes in India: political capital and sustainable livelihoods analysis', *Natural Resources Perspectives* 68 (London: ODI), [website] http://www.odi. org.uk/nrp/68.pdf, accessed 1 March 2008.

Beall, J. (1997), 'Thoughts on Poverty from a South Asian Rubbish Dump', *Institute of Development Studies* 28(3), 73–89.

Bello, W. (1993), 'Global economic counterrevolution. The dynamics of impoverishment and marginalisation', in Hofrichter, R. and Gibbs, L. (eds), *Toxic Struggles. The Theory and Practice of Environmental Justice* (Philadelphia: New Society Publishers).

Boonyabancha, S. (2005), 'Baan Mankong: Going to scale with "slum" and squatter upgrading in Thailand', *Environment and Urbanization* 17(1), 21–46.

Booth, D., Holland, J., Hentschel, J., Lanjouw, P. and Herbert, A. (1998), *Participation and Combined Methods in African Poverty Assessment: Renewing the Agenda*, Department for International Development, Social Development Division, African Division, London.

Brazil (2001), *Estatuto da Cidade*. Guia para implementação pelos municípios e cidadãos. Série fontes de referência. Legislação No. 40, Câmara dos Deputados, (Brasília: Governo do Brasil).

Brocklesby, M.A. and Fisher, E. (2003), 'Community development in sustainable livelihoods approaches – an introduction', *Community Development Journal*, 38(3), 185–98.

Brohman, J. (1996), *Popular Development Rethinking the Theory and Practice of Development* (Cambridge: Blackwell).

—— (2000), 'Popular development', in Corbridge, S. (ed.), *Development: Critical Concepts in the Social Sciences*, Volume V: Identities, Representations, Alternatives (London: Routledge). pp. 245–72.

Bromley, R. (1997), 'Working in the streets of Cali, Colombia: survival strategy, necessity, or unavoidable evil?', in Gugler, J. (ed.), *Cities in the Developing World* (Oxford: Oxford University Press).

Bury, J. (2004), 'Livelihood in transition: transnational gold mining operations and local change in Cajamarca, Peru', *The Geographical Journal* 170(1), 78–91.

Caldeira, T.P. (2000), *City of Walls: Crime, Segregation and Citizenship in São Paulo* (Berkeley: University of California Press).

Capobianco, J.P.R. and Whatley, M. (2002), *Billings 2000 Ameaças e Perspectivas para o Maior Reservatório de Água da Região Metropolitana de São Paulo* (São Paulo: Instituto Sócioambiental).

Carley, M. (2001), 'Top-down and bottom-up: the challenge of cities in the new century', in Carley, M., Jenkins, P. and Smith, H. (eds), *Urban Development and Civil Society* (London: Earthscan).

Carney, D. (1998), 'Implementing the sustainable rural livelihoods approach', in Carney, D. (ed.), *Sustainable Rural Livelihoods: What Contribution can we Make?* (London: Department for International Development). pp. 3–23.

Chambers, R. (1989), 'Vulnerability, coping and policy', *IDS Bulletin* 20(2), 1–7.

Chambers, R. and Conway, G.R. (1992), *Sustainable Rural Livelihoods: Practical Concepts for the 21st Century* (London: IDS).

Chambers, R. and Guijt, I. (1995), 'DPR: después de cinco anos, en qué estamos ahora?', *Bosques, Arboles y Comunidades Rurales* 26, 4–15.

Chant, S. (2004), 'Urban livelihoods, employment and gender', in Gwynne, R. and Kay, C. (eds), *Latin America Transformed: Globalization and Modernity* (2nd edn) (London: Arnold).

Chen, S. and Ravallion, M. (2007), *Absolute Poverty Measures for the Developing World, 1981–2004*, World Bank Policy Research Working Paper 4211, April 2007.

City of Seattle (1993), *The Sustainable Seattle 1993 Indicators of Sustainable Community*. A report to citizens on long-term trends in our community, Seattle City Council.

Conway, T. (ed.) (2001), 'Case studies on livelihood security, human rights and sustainable development'. Paper presented at the *Workshop on Human Rights, Assets and Livelihood Security, and Sustainable Development*, London Bridge Hotel, 19–20 June 2001 (London: ODI), [website] http://www.odi.org.uk/pppg/ tcor_case_study.pdf, accessed 1 March 2008.

Conway, T., Moser, C., Norton, A. and Farrington, J. (2002), 'Rights and Livelihoods Approaches: Exploring Policy Dimensions', *Natural Resource Perspectives* 78, May.

CSJ (Commission on Social Justice) (1998), 'What is social justice?', in Franklin, J. *Social Policy and Social Justice: The IPPR Reader* (Cambridge: Polity Press).

Davies, B.P. (1968), *Social Needs and Resources in Local Services* (London: Michael Joseph).

DFID, EC, UNDP and The World Bank (n.d.), 'Linking poverty reduction and environmental management: Policy challenges and opportunities', [website] http://wbweb4.worldbank.org/nars/eworkspace/ews004/groupware/GI_View. asp?ID=29, accessed 5 June 2004.

Dicken, P. and Lloyd, P.E. (1981), *Modern Western Society* (London: Harper & Row).

Dovie, D.B.K. (2002), 'Towards Rio +10 – trend of environmentalism and implications for sustainable livelihoods in the 21st century, the context of southern African region', *Environment, Development and Sustainability* 4, 51–67.

Drakakis-Smith, D. (1996), 'Third World Cities: Sustainable Urban Development II – Population, Labour and Poverty', *Urban Studies* 33, 673–701.

ECLAC (Economic Commission for Latin America and the Carribbean) (2007), *Social Panorama of Latin America 2007*, ECLAC: Social Development Division and Statistics and Economic Projections Division (Santiago: United Nations).

Erdmann, G. (2002), 'Neo-patrimonial rule – transition to democracy has not succeeded', *Development and Cooperation* 1, 8–11.

Estado de São Paulo (1997), *Termo de Referência para a Recuperação Ambiental da Bacia Billings*, Secretaria do Meio Ambiente SEMA: São Paulo.

—— (1998), *Plano Emergencial de Recuperação dos Mananciais da Região Metropolitana de São Paulo*, Secretaria do Meio Ambiente SEMA: São Paulo.

Fainstein, N. (1996), 'A note on interpreting American poverty', in Mingione, E. (ed.), *Urban Poverty and Underclass* (Cambridge: Blackwell Publishers).

Farrington, J., Tamsin, R. and Walker, J. (2002), *Sustainable Livelihoods Approaches in Urban Areas: General Lessons, with Illustrations from Indian Cases* (ODI Working Paper 162) (London: Overseas Development Institute).

Fernandes, E. and Rolnik, R. (1998), 'Law and urban change in Brazil', in Fernandes, E. and Varley, A. (eds), *Illegal Cities. Law and Urban Change in Developing Countries* (London: Zed Books).

Fix, M. (1996), 'O Estado e o capital nas margens do rio Pinheiros, duas intervenções: Faria Lima e Água Espraiada' (Tese de Mestrado, Faculdade de Arquitetura e Urbanismo, Universidade de São Paulo).

Freire, P. (1970), *Pedagogy of the Oppressed* (New York: Continuum).

Gans, H.J. (1996), 'From "Underclass" to "Undercaste": Some observations about the future of post-industrial economy and its major victims', in Mingione, E. (ed.), *Urban Poverty and Underclass* (Cambridge: Blackwell Publishers).

HABITAT (United Nations Centre for Human Settlements) (1982), *Survey of Slum and Squatter Settlements* (Dublin: Tycooly International Publishing Limited).

Harvey, D. (2001), *Spaces of Capital: Towards a Critical Geography* (Edinburgh: Edinburgh University Press).

Hinshelwood, E. (2003), 'Making friends with the sustainable livelihoods framework', *Community Development Journal* 38(3), 243–54.

Hjorth, P. (2003), 'Knowledge, development and management for urban poverty alleviation', *Habitat International* 27, 381–92.

IBGE [Instituto Brasileiro De Geografia e Estatisica] (n.d.), Contagem da População 1996, 2000 (Census data 1996, 2000), São Paulo: Fundação Instituto Brasileiro de Geografia e Estatística, [website] http://www.ibge.gov.br, accessed 17 October 2005.

International Labour Organization (ILO) (1996), *Social Exclusion and Anti-poverty Strategies*, International Institute for Labour Studies, Geneva.

Kirby, A.M. and Pinch, S.P. (1983), 'Territorial justice and service allocation', in Pacione, M. *Progress in Urban Geography* (Totowa: Barnes and Noble Books), pp. 223–50.

Kowarick, L. (1988), *As Lutes Sociais e a Cidade: São Paulo Passado e Presente* (São Paulo: Paz e Terra).

Lavinas, L. and Nabuco, M.R. (1995), 'Economic crisis and tertiarization in Brazil's metropolitan labour market', *International Journal of Urban and Regional Research* 19(3), 358–68.

Low, N.P. and Gleeson, B.J. (2001), *Justice, Society and Nature. An Explanation of Political Ecology* (London: Routledge).

Mahon, R., Almerigi, S., McConney, P., Parker, C. and Brewster, L. (2003), 'Participatory methodology used for sea urchin co-management in Barbados', *Ocean and Coastal Management* 46, 1–25.

Mazzucchelli, S.A. (1995), 'Participatory methodologies for rapid urban environmental diagnoses', *Environment and Urbanization* 7(2), 219–26.

Meikle, S. (2002), 'The urban context and poor people', in Rakodi, C. (ed.), *Urban Livelihoods: A People-centred Approach to Reducing Poverty* (London: Earthscan), pp. 37–51.

Meneguello, C. (2002), 'Conservation of City Centres' notes on the case of São Paulo, Brazil UNESCO Virtual Congress, October–November 2002.

Meyer, M., Hyde, M.M. and Jenkins, C. (2005), 'Measuring sense of community: A view from the streets', *Journal of Health and Social Policy* 20(4), 31–50.

Mingione, E. (1996), 'Urban poverty in the advanced industrial world: Concepts, analysis and debates', in Mingione, E. (ed.), *Urban Poverty and Underclass* (Cambridge: Blackwell Publishers).

Miranda, L.C.S. (2007), Vizinhos do inconformismo: o movimento dos sem-teto de Salvador. Salvador (UFBA), [website] http://www.uel.br/grupo-pesquisa/gepal/segundogepal/LUIZ%20CEZAR%20DOS%20SANTOS%20MIRANDA, accessed 20 April 2007.

Mitlin, D. (2003), 'Addressing urban poverty through strengthening assets', *Habitat International* 27, 393–406.

Moser, C. (1998), 'The asset vulnerability framework: Reassessing urban poverty reduction strategies', *World Development* 21(1), 1–19.

Narayan, D., Chambers, R., Shah, M.K. and Petesch, P. (2000), *Voices of the Poor; Cying Out for Change* (Oxford: Oxford University Press).

Potter, R.B. and Lloyd-Evans, S. (1998), *The City in the Developing World* (Harlow: Longman).

Prefeitura do Município de São Paulo (2000), *São Paulo em Números*, São Paulo: Prefeitura de São Paulo, Secretaria Municipal do Planejamento.

Rakodi, C. (2003), 'A livelihoods approach – Conceptual issues and definitions', in Rakodi, C. (ed.), *Urban Livelihoods: A People Centred Approach to Development* (London: Earthscan), pp. 1–22

Rakodi, C. and Lloyd-Jones, T. (2002), *Urban Livelihoods – A People-centred Approach to Reducing Poverty* (London: Earthscan).

Room, G. (ed.) (1995), *Beyond the Threshold. The Measurement and Analysis of Social Exclusion* (Bristol: The Policy Press).

Roseland, M. (1998), *Toward Sustainable Communities: Resources for Citizens and their Governments* (Gabriola Island: New Society Publishers).

Rouse, J. and Ali, M. (2001), 'Waste Pickers in Dhaka', *Water, Engineering and Development Centre*, (Loughborough University).

SABESP (Companhia de Saneamento Básico do Estado De São Paulo) (1997), *Água um bem limitado*, in ÁGUA – Associação Guardiã da Água – 2004, [website] http://www.agua.bio.br/botao_d_L.htm, accessed 21 October 2005.

Sakai, M. (2002), 'Enabling self-help activities through loan services in Thailand: The urban community development office's strategies for low-income community improvement', *Regional Development Dialogue* 23:1, 136–56.

Saraví, G.A. (2004), 'Urban segregation and public space: young people in enclaves of structural poverty', United Nations, *CEPAL Review* 83, 31–46.

Satterthwaite, D. (2001), 'Reducing urban poverty: constraints in the effectiveness of aid agencies and development banks and some suggestions for change', *Environment and Urbanization* 13(1), 137–57.

Sen, A.K. (1981), *Poverty and Famines: An Essay on Entitlements and Deprivation* (Oxford: Oxford University Press).

—— (1992), *Inequality Re-examined* (Oxford: Clarendon Press).

Soto, de H. (1989), *The Other Path: The Invisible Revolution in the Third World* (New York: Harper & Row).

Taschner, S.P. (1995), 'Squatter settlements and slums in Brazil: twenty years of research and policy', in Aldrich, B.C. and Sandhu, R. (eds), *Housing the Urban Poor, Policy and Practice in Developing Countries* (London: Zed Books).

Torres, H., Alves, H. and Oliveira, M. de A. (2005), *São Paulo Peri-Urban Dynamics: Some Social Causes and Environmental Consequences*, [website] http://www.worldbank.org/urban/symposium2005/papers/torres.pdf, accessed 3 February 2006.

UN-HABITAT (United Nations Human Settlement Programme) (2003), *The Challenge of Slums: Global Report on Human Settlements* UN-Habitat, (London: Earthscan).

—— (2006), *The State of the World's Cities Report 2006/2007* (London: Earthscan).

UNDP (United Nations Development Programme) (2003), *Human Development Report 2003* (New York: Oxford University Press).

United Nations (2005), UN Millennium Development Goals, [website] http://www.un.org/millenniumgoals/goals.html, accessed 17 November 2007.

Valladares, L. do P. (2005), *A Invenção da Favela: Do Mito de Origem a Favela.com* (Rio de Janeiro: Editora FGV).

Walker, R. (1995), 'The dynamics of poverty and social exclusion', in Room, G. (ed.), *Beyond the Threshold: The Measurement and Analysis of Social Exclusion* (Bristol: The Policy Press).

Yapa, L. (1998), 'The poverty discourse and the poor in Sri Lanka', *Transactions of the Institute of British Geographers* 25, 95–115.

Young, I.M. (1990), *Justice and the Politics of Difference* (Princeton: Princeton University Press).

—— (1994), *Justice and the Politics of Difference* (Princeton: Princeton University Press).

Zanen, S.M. and De Groot, W.T. (1991), 'Enhancing participation of local people: some basic principles and an example from Burkina Faso', *Landscape and Urban Planning* 20, 151–8.

Chapter 4

Grassroots Resource Recovery
for Human Security[1]

Introduction: Local Community Development and Bottom-up Initiatives

Collective recycling programmes are expanding, particularly in developing countries. In this chapter I will discuss one experience of community-based recycling in the periphery of São Paulo. The story touches on different frameworks – from local community development (Maser 1997) to grassroots initiatives (Mitlin 2003). The concepts of empowerment, social capital and participation are relevant in this context in order to find answers to the following questions: Is community-based recycling an effective form of resource recovery, promoting sustainable lifestyles? What are the constraints and the assets of such community initiatives? Answers to these questions can contribute to the process of building more sustainable communities. Analyzing local experiences are ways to add to what we know about building sustainable communities.

The terms *neighbourhood* and *community* will be used as interchangeable terms, based on the perception of the local residents. The underlying concepts are interchangeable, representing more than just a group of neighbours or a district. The terms imply a locality where people share similar characteristics or experiences. In our case it is the experiences of deprivation and exclusion but also solidarity and collective action (see also Chapter 3). Neighbourhood boundaries are flexible, and primarily follow the history of occupation and the income distribution and socio-cultural background of the residents. Grassroots initiatives are growing as economic and environmental conditions deteriorate for the poor, and as both state and local government are failing to respond to the unsatisfied local needs.

Community: The New Action Space

Community – determined as a group of people with similar interests and shared locality – can contribute in various different ways to local development. Shared membership is what makes a community. It implies in a place (sometimes virtual) where people meet. Local community development is a process of participation and action that shapes the community in the way its active members want it to be shaped. Young et al. (2004) talk about sense of neighbourhood, defined as having trustful

1   Part of the data presented in this case study was first published in 2002 in the chapter with Takahashi, M.R. in: Botta, H. Berdier, C. and Deleuil, J.-M. (eds), *Enjeux de la Propreté Urbaine*. (Lausanne: Presses Polytechniques et Universitaires Romandes), pp. 103–121.

social relations in the neighbourhood, which contributes to a feeling of cohesion and attachment to this community. Community involvement can then become a conscious process of self-determination (Maser 1997). Community members get empowered through the shared experience of collective action and the sense of accomplishment that flows from that action. Luckin and Sharp introduce the concept of *social capital* as a measure of the '…extent of social networks and norms of trustworthiness and reciprocity that exist in a community' (Luckin and Sharp 2006, 63).

Although stakeholder participation in public decision-making is a challenging endeavour, it is changing governance towards what Roseland calls the *politics of inclusion* (1989, 182), allowing strong communities to emerge in the process. Today, *good governance* means allowing for public participation and inclusive decision-making. Governance comprises the traditions, institutions and processes that determine how power is exercised, how citizens are given a voice, and how decisions are made on issues of public concern (Institute on Governance, 2007). Although not yet widespread reality, there is growing acceptance that participatory approaches ensure better results and have a higher likelihood of reflecting public interests than decisions achieved through public consultation or government lobbying. Active *authentic* participation in the decision-making process is essential (Timbo 2003). Active and authentic is here contrasted with passive and manipulated. This translates into citizens' participation in defining priorities, finding adequate solutions to problems and deciding what actions to take. Taking the community as a major focus for action is also a way of valuing the so-called *third sector*, which consists of independent, voluntary and non-profit organizations. Community is a place where people encounter each other, often through face-to-face daily interactions. The level of organization within the community usually shapes the local scale of action.

Local community development '…is a process of organization, facilitation, and action that allows people to create a community in which they want to live through a conscious process of self-determination' (Maser 1997, 101). The most innovative aspect of local community development is that, through mutual efforts to resolve shared problems, people increase their ability to control their own lives, thereby creating more fulfilling livelihoods for themselves. This approach asserts that, through collective action, people can successfully implement change. However, when addressing local development, the social contradictions, differences and conflicts that exist on the community level become evident. Participatory strategies based on strong communication can allow partners to overcome differences among themselves through empowerment, rule making, conflict management, power sharing, social learning, and open dialogue.

There are numerous documented – and probably even more undocumented – examples from all over the world where people have made a difference to the quality of life in their neighbourhoods by initiating recycling schemes. Community-based recycling initiatives are on the rise. In the South these initiatives are driven both by individual's needing to generate income and communities lacking official waste collection; in the North the initiatives are motivated by the elevated cost of regular waste management and the social and environmental benefits of resource recovery. In the North these experiences, which often date back to the early 1970s, are concentrated in smaller, progressive communities. Many community development experiences in the North deserve to be highlighted, such as those on the Hornby, Saltspring, and Mayne Islands in Western Canada; the *Emilia Romana* region in Northern Italy; and the Community Recycling Network UK in Bristol, UK. In the South, selective collection of recycling materials is providing a livelihood for many of the urban poor in almost every country. Particularly in deprived neighbourhoods in metropolitan areas, grassroots initiatives often address these issues with innovative community-based solutions (Besen 2006).

These communities in the South are not only poor; they are also deprived of many necessities, not just the pressing economic ones. The concept of social exclusion helps explain why so many people are poor, why they remain in poverty and how the cycle of poverty can be disrupted. In this chapter I will discuss promoting participation in recycling as a possible strategy to improve human security. This is an example of pursuing a bottom-up approach in local development, aiming at building sustainable communities. While, in theory, bringing the affected parties into the decision-making process seems necessary, it is in practice often the most difficult goal to achieve. This story speaks about the successes and failures of a bottom-up development experience, and showcases some of the difficulties the community had to face while implementing this project.

A new urban social movement has emerged in Brazil, involving the informal independent recyclers and the waste recovery sector organized in recycling associations, community groups and co-operatives. These groups and individuals collect recyclables in the street, at the household level or at firms. One important grassroots recycling network is the Movimento Nacional dos Catadores, an alliance of the many organized selective waste collection and recycling initiatives. This national recyclers movement is a vibrant social movement in Brazil, and over the past years it is gaining political importance together with the other two social movements: the landless (*Movimento dos Sem Terra*) and the homeless movement (*Movimento Sem Teto*), the latter one was already introduced in Chapter 3. The recyclers have already held three international congresses, besides many regional and national meetings, demonstrations and government negotiations in many cities in Brazil. There are also municipal networks, such as the *Forum Recicla São Paulo* (Recycling Forum in São Paulo), which aggregates more than 30 recycling initiatives in the city of São Paulo. The purpose of this forum is to empower the member groups. In a participatory governance structure that includes these recycling initiatives in local waste management, organized forms of this category like the recyclers' forum play an important role in terms of enhancing the activity and improving the livelihoods of the recyclers by lobbying for the introduction of specific policies that target income generation, health, education and social inclusion.

Painter (1995) defines social movements as '…groups of people acting collectively in pursuit of shared goals, which include, or require, social and/or political change' (153). Collective action can become a powerful strategy to promote positive change, in the sense of sustainability. One of the reasons for the emergence of social movements, noted by Camilleeri and Falk, is '…the failure of the state to resolve the contradictions between economic growth and development' (cited in Painter 1995, 156). Collective action occurs for specific reasons. Collective identity is a central concept in social movement theory. Shriver et al. (2000) affirm that shared group definition increases the likelihood that individuals will participate in a movement. 'Group members' common interests and experiences enhance group solidarity, generating shared beliefs and norms that underlie the emotions supporting activism' (Shriver et al. 2000, 43). This is exactly what happened in Brazil, as informal and organized recyclers rose into a new social movement. The obvious opportunity for recycling to become a strategy for redistribution of income and for poverty mitigation is probably the most important political aspect of this movement. The driving forces and the weaknesses involved in this endeavour will be discussed.

The case study is located in São Paulo, in the neighbourhood *Pedra sobre Pedra*, which has already been introduced in the previous chapters. Here, the benefits and hurdles of communal recycling in the context of a developing country will be considered. A few leaders of the neighbourhood association have introduced recycling and environmental education as instruments to improve the quality of life in their community. The social assets these groups bring and the different political, economic and social barriers they have to face will be carefully considered during the evaluation of this experience. The chapter concludes with an examination of other bottom-up recycling activities in developing countries, linking back to the global perspective on the potential benefits these initiatives can bring to local communities.

People's Development Initiatives and the Power from Below

Current people-centred development thinking conceptualizes participation, empowerment and capacity-building as fundamental for social transformation. Empowerment is strengthened by decentralization, transparency and accountability in all aspects of governance, including the management of natural resources (UNFPA 2001).

Through structural interpretation, Timbo (2003, 21) defines social transformation as '...a process of change that results from continuous action by agents on the different forms of structures that they face, within and outside of society'. People are agents of change, however, the existing structures often inhibit the aspired change of the people. Whether individuals within a society can transform their society is effected by what is possible within that given social, historic and political setting. There are structures that limit the drive for transformation. Assets-based approaches focus on the idea of expanding people's capabilities and assets so that people can become stronger agents for change (Foster and Mathie 2003).

From a sustainability perspective, people need to be involved in the decisions that affect their lives and reflect their interests. The process is one of social transformation – challenging oppressive rules and norms of the social, economic and political system and engaging with the existing impediments to improve livelihood situations. Hence, the focus is on those that are affected by unjust structures and those who perpetuate them. Deconstructing specific local and micro contexts of a situation assists in understanding social change and the links to a global scale. Focusing on the local level or the people's livelihoods helps us understand the larger context of development.

Participation in itself can be part of the people's learning process. This involves a shift in power and a challenge of the prevailing power structures. Empowerment means the enhancement of human freedom. Empowerment is an indicator of and a motivating force behind social development (Sen 1999). Giddens (1991) defines it as '...the power of human beings to alter the material world and transform the conditions of their own actions' (cited in Simon 2003, 5). Empowerment has also been defined as a process '...by which people, organizations and groups who are powerless become aware of the power dynamics at work in their life context, develop the skills and capacity for gaining some reasonable control over their lives, exercise this control without infringing upon the rights of others and support the empowerment of others in the community' (cited in Rowlands 1995, 103).

In other words empowerment is about structural transformation, and it means to become aware of hidden power structures and learning the tools that enable social change and applying them. It is about changing '*power over*' relationships by giving a voice and power to the powerless to pursue their interests. Participation and empowerment can tackle inequalities and imbalances in access and distribution of resources for development. Questions of justice depend on power, for example the power to set rules and define norms (Timbo 2003).

Following Giddens' *Structuration Theory*, action is affected by structural contexts and also by agency (*duality of structure*). The people who compose society, their relations, institutions and events all influence action and are also influenced or even constrained by it. 'Action is a product of people's agency and of the rules and resources of the particular social system in which they live' (Panelli 2004, 189).

According to Amartya Sen (1996) it is not only the unequal distribution of commodities which is problematic, but also the resulting unequal capabilities of people, deriving from a situation where only some have access to goods and services or to a certain quality level of these goods and services. There is a clear North/South divide in terms of the level of inclusion/exclusion. Social exclusion has become a prevalent component in the South where wealth is, to a large extent, based on the exploitation of others and/or the environment. The term explains a situation that goes beyond poverty and consists of the separation of individuals or groups from the rest of society through economic deprivation as well as social and cultural segregation. As already highlighted in Chapter 3 a significant proportion of the population in developing countries is socially excluded.

Often structural problems perpetuate exclusion, such as for example, in consequence of not having a postal address people are unable to open a bank account and can not access most formal social programmes. Linear, deterministic and homogenizing interpretations of poverty and exclusion do not explain the complex livelihood situations found in poor neighbourhoods. Community development focusing on human agency, continuous learning and recognition of diversity, can better elucidate and address these issues. Approaches 'from below' need to be widely discussed, not only in the academic forums but also in the governmental sphere. Opportunities need to be created to put in practice innovative forms of collective action, in order to make a new pathway for change.

Building Social Capital with Recycling

Community-based collective processes can increase social cohesion and enhance social capital, which builds stronger communities. It seems awkward to translate social dynamics and social values into monetary values. Furthermore, economic progress is not the only and maybe not the most important indicator for development. Yet, the concept of social capital can contribute to a better recognition of the importance of social conditions, in making evident its contribution to economic development. It provides a way to conceptualize an important source of capacity present in the community: the resource potential of personal and organizational networks (Chaskin et al. 2006). Social capital looks at the role of collective, community-oriented actions, trust and reciprocity; all aspects that promote better governance. Improved livelihoods, safer neighbourhoods, responsive governance, less environmental impacts and overall healthier communities are some of the positive outcomes from the presence of social capital in the community.

Chaskin and the other authors emphasize '...the role (for good or ill) that social structure – the concrete networks of relations among individuals and institutions that define the shape of social interaction – plays in providing access to information, opportunity, and support. It is through these relationships, and the associational action they make possible, that social capital operates' (2006, 492). Encouraging and facilitating the creation of social networks, can therefore be a pro-active form of capacity building, aiming at expanding social capital. There is the potential to build shared values, achieve a more equitable distribution of and access to resources, reduce income disparities, and empower people as being part of a common project and as members of the same community. These are active forms of expanding social capital. Community engagement also refines one's 'sense of community', determined by '...the extent to which community members experience a sense of solidarity and a sense of significance' (Young et al. 2004, 2628).

Volunteer involvement strengthens citizenship in the community and builds social capital (Luckin and Sharp 2003, 2004, 2005). It generates opportunities for individuals to gain confidence and to develop skills and expertise that benefit the individual for example, in terms of human development or professional training. Community-based waste management also opens possibilities for the development of relationships in the larger community, strengthening the social networks. This collective exercise provides the ground for inclusive decision-making and civic engagement (Muller et al. 2002). Participation in recycling schemes demands a personal commitment that challenges the routines of a throwaway society, and it is argued that recycling can lead to other environmentally beneficial changes in individual behaviour.

Informal Recycling Generating Income

We have already seen how important informal recycling is for poor people to generate income. Recycling is a widespread activity, not only in developing countries, but in countries with large income disparities like the USA, Canada and countries of the former Soviet Union. In Canada or the USA recyclers are known as binners, in Brazil as *catadores* (pickers) or *carrinheiros* (pushcart users), in Argentina as *cartoneros*, *classificadores* or *recuperadores* and more generally as *recicladores*, the Portuguese and Spanish term for *recyclers* (see Illustration 2.2). An estimated 200,000 informal recyclers work in Brazil. According to a countrywide survey conducted by the NGO network Fórum Nacional Lixo e Cidadania, 37 per cent of the municipalities in Brazil have acknowledged the existance of informal recyclers separating at landfills, particularly in cities over 50,000 inhabitants (Programa Nacional Lixo e Cidadania, n.d.). 67 per cent of the capital cities confirmed and only 11 per cent denied the presence of informal recyclers working in the streets. Most of the street recycling happens in larger cities. According to this study, municipal governments have recognized approximately 45,000 full-time informal recyclers on landfills and 30,000 in the streets of the major cities in Brazil. However, the authors recognize that these numbers might be widely underestimated (Programa Nacional Lixo e Cidadania n.d.). In 2000, CEMPRE, the industry sponsored, not for profit organization, that promotes recycling as part of integrated waste management conducted a different census of informal recyclers only in the city of São Paulo. They identified 3,686 recyclers, of whom 448 were children (SBPC 2002), a number that according to other sources Romani (2004) seemed to be underrated. Recent data (Romani 2004) on other cities in Brazil provide the following approximate picture for the year 2004: 5,000 recyclers in the city of Belo Horizonte, 3,000 recyclers in Porto Alegre, 2,000 recyclers in Recife, 2,000 recyclers in Rio de Janeiro and 376 recyclers in the ABC municipalities of the metropolitan region of São Paulo. Again, it is expected that the actual numbers are much larger, since a comprehensive census has not yet been conducted (Besen 2006). Because recycling is an informal activity, it is difficult to obtain exact numbers; and since this is a volatile sector, closely dependent on economic performance indicators, the figures fluctuate significantly over time.

Child labour in resource recovery in the streets and at landfills is obviously a prohibitive situation. Since 1998, preventing children from working with solid waste has become one of the major targets for UNICEF in Brazil, which by then estimated a total of 45,000 children working in this sector. In 1999, the Fórum Nacional Lixo e Cidadania started the campaign *Criança no Lixo Nunca Mais*, which helped significantly diminish the number of children working at landfills (Programa Nacional Lixo e Cidadania n.d.). The factors that lead to child labour are multifarious, and are primarily related to poverty. A recurrent problem for women recyclers is where to leave their children while working. Many recyclers say that they prefer to work in their children's company rather than have them unwatched in the streets or locked at home. Still today the alternative for many women involved in recycling is to either leave the children with other family members or bring them to work. This issue could be more readily addressed with co-operative working schemes. ASMARE – the leading recycling co-operative in Belo Horizonte, with 380 participants in 2006 – offers free child care for its members. Appropriate policies and projects to protect children and childhood are needed to address the pressing social, cultural and economic issues that arise from social exclusion.

The size of the informal recycling sector indicates the extent of poverty. Selective collection of waste is the last resort of income generation for vulnerable, disempowered and socially excluded individuals willing to do honourable work. The academic literature gives evidence of a close association between poverty and social exclusion (Silver 1994; Room 1999; Waggle 2002; Legros 2004). In order to tackle poverty it is important to understand it and the underlying concept of social exclusion (Brady 2003). It is noticeable that urban concentration of poverty is growing on a global scale (Hjorth 2003), resulting in an increasing number of people excluded from the formal economy (Beall 2000).

Resource recovery can also be an indicator for environmental sustainability. The more organized and structured resource recovery happens, the greater the sustainability of its community. For the *catadores*, waste is conceived as a resource that has value when being recovered and as such for them it is an important means of income generation. Their work consists of selectively collecting recyclables from businesses, schools, apartment buildings, or out of the garbage in the streets and sometimes at landfills, and transporting them for separation at home, at recycling centres, or at the middlemen's premises. Recyclers may walk more than 30 km a day while collecting, and they usually work long days, often seven days a week (Cazetta 2005). Handling and processing recyclable materials exposes the workers on a daily basis to dangerous and unhealthy conditions. The health aspects of informal recycling will be discussed in Chapter 5.

The informal recyclers remove the most valuable resources (aluminium cans, glass, plastic bottles and paper) out of garbage bins or plastic bags placed in the street, before the official waste collection. This practice, which the recyclers call *Pente fino*, often means scavenging through filthy waste. The public generally disapproves the activity because it may involve littering. With an organized selective collection, this negative side effect can be eliminated.

Independent recyclers generally separate in the street or at home and sell their proceeds on a daily basis to middlemen. Organized recyclers usually classify at triage centres or specific spaces provided by co-ops, associations or the municipality. Sometimes they are equipped with a balance and press, and a few groups even have a truck for the collection and commercialization of the material. The infrastructure for separation varies significantly and in most cases does not correspond to ergonometric requirements. Very few of the independent and organized recyclers wear gloves or mouth protection during the collection and separation.

Independent recyclers earn on average 1/3 or less of what the industry pays to waste dealers. The recyclers are part of a cycle of exploitation unless they join larger groups (co-operatives or associations) and connect with networks for commercialization. According to Medina (2000) middlemen and waste dealers can achieve high profits because they operate in a *monopsonistic* market, dominated by only one buyer, as opposed to *monopolic* markets with only one seller. Under these circumstances the prices are dictated by middlemen, who usually pay less than under more competitive market conditions. Here is a real opportunity to add value to the work of the recyclers: setting up networks to improve the logistics and strengthen the shared commercialization.

Community – Government Partnerships in Recycling

Recycling has become an official waste management option in many cities in Brazil, spearheaded by Londrina, Belo Horizonte, Porto Alegre, Itú, Curitiba, Diadema and Embu. Sometimes local governments provide specific incentives for recycling. The case of Curitiba (1,587,315 inhabitants) in the south of Brazil is also well known for its *Purchase of Garbage Program*, launched early in 1990. The programme involved more than 22,000 families from low-income households. Participants could hand in their bags of garbage in return for bus tickets and agricultural and dairy products. Marginalized neighbourhoods, which had been chronically suffering from irregular waste deposits, became cleaner. Today almost half of all the rigid plastic is recovered out of the domestic waste stream in this city (CEMPRE 2000).

Since 1993, ASMARE has been involved in a formal partnership with the local government in Belo Horizonte, through the city's agency for sanitation (SLU). This has allowed membership at ASMARE to increase from 11 to 380 members. Operational infrastructure and technical support have been part of an agreement between the government and the association. The policy has contributed to an increase in resource recovery and has generated income for the poorest segment in the city.

With the rise in the prices paid for recyclables and with an increased number of impoverished and unemployed, the competition for resources discarded in the garbage has increased. At the same time, with growing consumption of highly packaged products, there is now more material available than ever before. Despite the urgency in adopting resource recovery and in tackling social and economic exclusion, few governments and communities have actively taken measures to address these pressing concerns.

Innovative pilot programmes, where the community has become a motor in recycling, are still scarce; however, their numbers are increasing. These projects contribute to increased environmental awareness and social cohesion, since the action has to come from the individual level, and they reinforce the sense of community of those involved in collective experiences. However, most cases remind us that a widely acepted paradigm shift away from the wasteful and resource intense disposal society has yet to happen.

On *Hornby Island*, on the West Coast of Canada, for example, a community initiative is engaged in the reduction, reuse, and recycling of household waste. Since the closure of the local landfill and the opening of the recycling depot in 1978, residents from this island have generated less than 0.5 kg of household waste per day, which is much lower than the provincial average. Here most of the waste is recycled. Hornby Waste Management Centre contributes to the reduction of waste management costs, and also generates permanent employment for three project managers (Hornby Recycles 2007). The facility consists of a free-store, user-pay garbage drop-off, composting site, composting toilet, and drought tolerant gardens. While the majority of the recycled material is sent off the island, part of it is also reused or transformed on the island by the islanders themselves. Glass has been used in handicraft, and in road and building construction. Thanks to the wide community engagement, in part founded in the sense of neighbourhood and the local social capital, Hornby Waste Management Centre is a success story (see Illustration 4.1).

Illustration 4.1   Hornby Island recycling depot, Canada

The municipality of Quito, Ecuador has expanded its waste collection through small-scale enterprises, created by and managed by residents in neighbourhoods that had no previous service provision. The revenues from the sale of recyclables go to a fund that supports neighbourhood improvements (Hernández et al. 1999). This pilot programme has contributed to income generation for the poor and has improved local environmental health. Similarly, in the city of Johannesburg in South Africa, locals involved in the collection of the waste now receive a salary from the city for their work. In this case, the municipality is still in charge of managing the solid waste, primarily with landfills (Swilling and Hutt 2001). Another example comes from the *Clean and Green Madras City Project*, an alliance between the public sector, an NGO, and the community in Chennai, India. This project provides the facility to educate 250 street children through their participation in recycling (Baud et al. 2001; see also Iyer 2001).

The example of organized recycling in the city of Londrina, in the south of Brazil, illustrates the quantity of resources that recycling groups divert. In 2005, approximately 500 *catadores*, organized in 26 groups, collected approximately 90 tons/day or 2,400 tons/month, according to the municipal administration of Londrina (Besen 2006, 117). On average each group consists of 18 members working eight hours a day. The amount collected varies from two to four tons of mixed material per day between the groups, depending on gender composition, motivation and level of expertise. In 2005 the average income per member was Reais $400 (US$164), which was above the minimum salary of Reais $300 (Besen 2006, 117).

Community recycling schemes are often the result of earlier local development initiatives, frequently implemented by non-governmental organizations or other outside organizations (Miraftab 2004; Robbins and Rowe 2002). They require financial and human resources and capacity, which is often not available in poor neighbourhoods. However, forming partnerships between multiple stakeholders from the civil society, business and government in order to address waste problems is one of the emerging alternatives to conventional waste programmes. There are good examples highlighting the benefits from private-public partnerships and community involvement in the UK (Luckin and Sharp 2004, 2005; Sharp and Luckin 2006).

Case Study: Neighbourhood Recycling a Bottom-up Experience

Neighbourhood associations have an important role to play in consolidating new urban spaces and providing basic public facilities. There were approximately 1,300 neighbourhood associations in São Paulo by 1989 (Durning 1989, 10). The neighbourhood association *Associação Pedra sobre Pedra* (APSP), already mentioned in Chapter 3, was created in 1991 to assist in the occupation and organisation of the new settlement. Jocemar Silveira was the key figure among a group of approximately ten other community members that began this initiative. Over the years, he became the squatters' most important voice in negotiating with the local government. With Jocemar at the forefront the association committed to several activities to improve quality of life and to increase environmental awareness among the local population.

Over more than ten years, Jocemar continues to be a fulltime community activist. He works on a variety of temporary contracts, just to provide for his livelihood but his true passion lies with environmental and social grassroots issues. Since then he has been involved in inspiring self-help projects, such as the invention of a hand-driven press for recyclables or the bio-digestor (he calls it *Redigestor*), which generates energy from organic household waste. Since the early 1990s his passion has been the recovery of recyclable materials and that's where this story starts as well.

In 1996, community members, motivated to change existing hazardous and unhealthy living conditions, mobilised to create a selective waste collection initiative. A few residents with leadership skills organized volunteer clean-ups in the settlement, promoted a workshop, and distributed information leaflets to stimulate environmental awareness. Most participants work in the informal sector, or have temporary, part-time jobs. The majority lives with some kind of economic hardship. Most of the participants did not mention political reasons for their involvement, though they might see obvious to an onlooker. Only Jocemar declared himself to be left wing, though without apparent political party orientation. However, he was also involved in some political campaigning in the past. The group networks with different local and regional politicians and had contacts to different non-governmental organisations.

APSP recognizes recycling also as a vehicle to draw public attention to the deprived living conditions in the neighbourhood, and its members believe that it was possible to gain public support for social inclusion through their activities. Most of the local residents who collaborated with the recycling proposal recognised the advantages of the project for their community. Despite the overall beneficial results and recognition from the community, the local government did not support the project and the working conditions continued to be precarious. The calamities that happened in 1999 during the summer rainfalls, with the consequent floods and landslides, have brought possible health threats and the risks related to irregular garbage disposal to the attention of the local population.

Initially the recycling project involved 15 members from the community (four women, six men and five male teenagers) who dedicated on average 16 hours/person/week to the project. Initially the recyclables were collected only at a few households and the population could bring their materials to the association, a small shed on a steep hillside with less than 50m² floor space. Volunteers placed large nylon bags (called *biggie-bages*), donated by a small-scale local enterprise, at central corners in the neighbourhood to collect the recyclables. Once a week APSP participants emptied the bags and separated the material at the shed, which is the only space of the association (see Illustration 4.2).

Until today these activities still take place, although some characteristics have changed. The project now collects recyclables from three schools, two small-scale businesses, four multi-apartment buildings and several households located outside the settlement. These materials are considered particularly valuable, both because they come cleaner and because there are more of them. The *door-to-door collection*, on the other hand, is regarded as *poor garbage*, and it garners only small amounts of valuable materials such as aluminium cans, white paper, or cardboard. Local informal recyclers and school recycling programmes contribute to the collection of

Illustration 4.2   *Biggie bage* **community recycling at** *Pedra sobre Pedra*

aluminium cans (for example, through the programmes supported by CEMPRE). The recyclables are transferred to the association, where they are separated into categories: aluminium, different kinds of metal and plastic (rigid plastics, plastic film, different coloured PET drinking bottles and other combined plastic materials), paper, cardboard, cardboard packaging (Tetra Pack), glass and others. Once classified, the material is compressed. At the time this group also separates approximately one ton of rejected material per month (for example, Isopor, wood, fabric, cloth, shoes) out of the *biggie bages*, which is then diverted to the formal waste collection.

The quantity of and prices for recycled material generally fluctuate, depending on the market and the quality of the material. Higher prices can be achieved for compressed material. Depending on the quantity of separated material, the buyer comes to pick up the load. There are many small-scale industries in São Paulo that depend on recycling initiatives like the one run by APSP. Table 4.1 shows the number and types of businesses involved with recycling in São Paulo, based on data collected by the city's waste management sector (Departamento de Limpeza Urbana).

The association's average gross income was Reais $800/month (US$440/month). However, they have to pay Reais $400 (US$220/month) monthly costs involving collection and separation (hourly wages and material costs). This amount does not include maintenance and eventual repair costs for the equipment. One of the aims of the project is to be able to provide proper wages to all people involved in the activity. Today, most of the work is still done on a voluntary basis. The participants see it as a worthwhile activity and are hoping to be able to make a living out of it

Table 4.1   Small-scale recycling business in São Paulo

Sector	Number of firms	Minimum amount accepted
Aluminium	7	0.5 to 7 tons
Metal	16	1 to 12 tons
Paper and Cardboard	23	1 to 10 tons
Plastics	7	5 tons
Glass	3	1 to 10 tons
Industrial residues	2	1 ton
Tissue sacs	1	500 items
Various: plastics, paper, metals, and so on	10	0.5 to 5 tons
TOTAL	69	
Sector specific associations supporting recycling in the region	13	

Source: Schneider, D. (2000), Cadastro de compradores de reciclaveis, LIMPURB (Departamento de Limpeza Urbana), Prefeitura do Município de São Paulo. São Paulo.

in the near future. According to Jocemar '...so far the only value from this story is that the packaging has not been carried into Lake Billings' (interview conducted in early 2000). Jocemar is clearly concerned with the environmental degradation in the periphery and with the waste problem of generating ever more waste.

The association promotes educational activities to increase environmental awareness, instructing the population about best practises in recycling and informing them about the outcomes: One of the members, a popular musician called Aleluia, has produced a CD with songs about the environment, in the hopes that the music will encourage people to recycle. Nevertheless, members of APSP admit that there is a general lack of public commitment and participation. Awareness and environmental stewardship are missing in the community, and it is apparent that resignation and scepticism towards any collective involvement is still widespread. People are tired of the constant promises made by the governments without ever fulfilling them, and suspect any organisation that offers hope.

The project was constrained by a shortage of funding and the precarious working conditions and infrastructure, including the lack of space. During the first years, the return from recycling was low, and often the leader had to pay from his pocket to keep the project alive. To address these issues, members met with representatives from four other neighbourhood associations in the region, in September 1999. The meeting resulted in changes in the form of conducting the collection and commercialisation of recyclables. The association realized that it was difficult, time consuming and expensive to maintain the collection with the *biggie bages*. Access to the settlement was tenuous and some areas were completely inaccessible during rainy days. Most of the material donated by the households was still coming mixed and dirty. Not all residents had a clear understanding about the purpose of the project and its benefits to the community and so some withdrew their support. APSP recognized that even more environmental education was needed in order to obtain better quality recyclables and to increase the number of participants.

The new approach resulted in the concentration and intensification of the door-to-door collection, in six streets and seven alleyways, covering a total of approximately 100 households. All residences in this area were visited to explain the objectives of the recycling programme. 90 per cent of the participants confirmed their commitment to the project. As a result, after two months APSP had increased the amount of recyclables recovered (Table 4.2). Although more time-intensive, the face-to-face contact and conversation with residents resulted in a higher participation rate than was obtained by distributing leaflets or addressing the population through loudspeakers.

I became aware of this grassroots recycling activity when conducting research in 1999 on quality of life in this neighbourhood as described in Chapter 3. Ruth Takahashi and myself were then the key protagonists in the preparation of a workshop on environmental education, involving local leaders, researchers and the community. Our objective was to draw the attention of the government and the community to the contaminating aspects of the waste, which is not collected and finally ends up in the Billings drinking water reservoir. Recycling was presented as a practical action to improve the local environment and the living conditions. As a consequence of the debates emerging from this event, the recycling network *Fórum Recicla São Paulo* was created. It is an association that integrates the informal recycling initiatives in

Table 4.2   Insight into the accounting of a small recycling project
**   *Pedra sobre Pedra* (October 1999 to May 2000)**

Material	Collection period			
	22 October – 30 December 1999		1 March – 31 May 2000	
	kg	Gross income (R$)	kg	Gross income (R$)
Cardboard	1,500	135.00	–	–
Mixed – paper and cardboard	6,833	615.00	9.930	1,485.20
Rigid plastics	1,300	185.00	2.378	262.92
Plastic film	1,500	172.00	–	–
PET – green and transparent	1,015	385.00	1.600	371.20
Tetra Pack	–	–	1.550	70.00
Glass	–	–	4.200	167
Aluminium	–	–	16	16.80
Metal	2,200	132.00	507	36.70
TOTAL	14,348	1,624.00 US$894	21,181	2,409.82 US$1,326

Source: *Associação Pedra sobre Pedra* (2000), *Cadastro de levantamento das atividades de coleta seletiva*, São Paulo.

São Paulo. The initiative started in early 2000 with 15 groups, and today the number has increased to 29. The main purpose of the network is to promote co-operation among those groups and to facilitate their involvement in the collection, separation and commercialization of recyclables. *Fórum Recicla São Paulo* is concerned about gaining better prices, expanding the geographical scope of recycling, improving working conditions for recyclers, building greater awareness so that the public will participate in recycling, and improving environmental conditions. In 2005, the Fórum received minimal financial support as part of the CIDA (Canadian International Development Agency) funded Participatory Sustainable Waste Management Project to conduct regular meetings to mobilize the organized recycling sector in São Paulo, a support that was decisive in the survival of the network at that time.

Over the past years new leadership has emerged from within *Pedra sobre Pedra*. The recycling project has empowered APSP and has benefited the local community, and there is the potential to expand on the success of this group. However, support from the local community, government and business community is needed for organisations like this to thrive. Chronic lack of funding and hardship among the participants still continue. The main question is: How to motivate more genuine grassroots participation? Creative initiatives using music, theatre or video as communication tools, for example, can generate more acceptance and support from the population.

Due to conflicting personalities and divergent political perspectives among the members of APSP, Selva de Pedra emerged, a new neighbourhood association involved only with local community recycling. Today in 2007 both groups loosely collaborate in separating the material. APSP has 11 members and Selva de Pedra, led by Rozenir Rodrigues Souza, known as Aninha, has nine members who work part-time in the local selective collection and separation. Aninha is also a seriously committed environmentalist who believes that the environment can be cleaned up with recycling and income for the poor can be generated from this activity. Although very poor, she dedicates a significant part of her time and resources to this cause. Her livelihood is barely met. She lives as a single mother of two children in a small shanty in *Pedra sobre Pedra* and since she doesn't have a regular income, there is often not enough money to pay for her bills (see Illustration 4.3). Nevertheless she dedicates her time to the small community recycling group.

Illustration 4.3   Housing in *Pedra sobre Pedra*

APSP has two trucks; but according to the leader of APSP, because the trucks are old one of them is always broken. They collect from 60 partners, some of whom are located in other neighbourhoods, including apartment buildings, housing complexes, the City's General Assembly building, and the energy company EMAE. The monthly earnings for the recyclers at APSP remain low, with approximately Reais $200–250 and at Selva de Pedra between Reais $80–150. Most families live below the poverty rate and their survival is guaranteed only with government subsidies, such as *bolsa família* (family income subsidy), *renda minima* (minimum income subsidy), *vale gás* (subsidy for cooking gas) or/and *bolsa escola* (student scholarships).

The leader of APSP has invented several useful tools, including a hand powered press, and a re-digester for organic-waste-generating gas that can be used for cooking. Lack of funding remains one of the major problems. It is very difficult for such initiatives to be self-sufficient without government support, for example through paying for the environmental service of recycling and by providing better working conditions. A pilot project supported by the University of São Paulo is planning to test the *redigestor*, which might provide future sources of funding for the project.

Resource mobilization theory explains the success and failure of social movements by focusing primarily on the ability of groups to mobilize resources, including financial, material, or symbolic resources, such as people's time (Painter 1995). It takes more than rational cost-benefit analysis to engage individuals in collective action. A common idea and a shared sense of identity are necessary to mobilize people. Kriesi, Koopmans, Duyvendak, and Giugni (1995) define collective identity from the European perspective as '...a sense of solidarity, and a political consciousness' (cited in Shriver et al. 2000, 46). A shared experience, over an extended period of time, in the context of a collective struggle can build group cohesion and awareness. Resource-dependent barriers (time, money, information) can inhibit the development of a sense of solidarity that would translate into collective action. Financial resources to cover basic needs and to improve basic infrastructure – resources that are critically important in getting social movements growing – are often nonexistent.

Conclusion: The Contribution of Recycling to Urban Sustainability

The present case study describes a community recycling initiative in a squatter settlement at the outskirts of São Paulo. For community initiatives to become successful, there needs to be local leadership and activism so that participants get motivated to implement collective actions. In the case of the neighbourhood association of *Pedra sobre Pedra*, APSP Jocemar first was the only key figure in this recycling project. His involvement with politics has driven some of the community members away from him and therefore a fraction of this movement has split into the new group, led by Aninha, which is also collecting and separating recyclables. The situation illustrates the multifarious difficulties squatter settlers face during the implementation of neighbourhood projects. Funding constraints and lack of support from the government are among the most crucial limiting factors. On the other hand, there is perseverance of leadership that has over the years helped to promote environmental awareness about recycling.

The negative effects of poverty and social exclusion have been described as conditions that destroy self-confidence and self-respect (Durning 1989). For these reasons it is often difficult to organise the masses of marginalized population in squatter settlements, caught in '...the traps of fatalism and division' (Durning 1989, 19). Unless there is strong leadership in the community to catalyse the population's will for change, grassroots organization becomes difficult. There is usually mistrust, scepticism and fear in these extremely precarious neighbourhoods as a result of social exclusion the residents deal with in their day-to-day lives. Paul Cardan describes the preference of withdrawal from action in the North as being a result of 'modern bureaucratic capitalism, [where]…the externally imposed "rationalisation" with the maintenance of exploitation as its objective, soon destroys the meaning of work and of all social activities. It leads to a massive destruction of responsibility… [and]…initiatives tend to disappear' (1974, 62). This is also true in the context of poor neighbourhoods in the South, where economic hardship and the daily struggle to improve livelihood limit the collective action of most individuals. The difficulties in establishing a dialogue for partnership with the government, and the formalities to overcome in attempting to develop such partnerships, are so wide-ranging that many give up. Formalizing community groups in order to access funding is a bureaucratic undertaking, and therefore a route most organizations don't seek. It is difficult for neighbourhood associations to have a voice and it is even harder to be heard. Often the poor are portrayed to the public by the media and by economic and political dominant groups as fear inducing. This '…surely serves to some extent to produce the very behaviours that are dreaded, while also increasing the likelihood that such groups will be victimised (through hate crimes and or official brutality) with relative impunity. Discourses of fear are potent. We need to deconstruct these discourses and provide counter-discourses…[and]…we don't hear about their fears, the fear of being hungry and not being able to provide for their family. Fears need to be communicated and negotiated' (Sandercock 2002, 217).

Social transformation, although a burdensome and laborious process, can only happen through popular participation. It depends on the extensive input of time and other resources of key leaders. Until now, too often the government has ignored the severity of the circumstances lived in the periphery, and it is only through the voices of the excluded that the reality is spelled out. Valdemir, a member of the neighbourhood association APSP, noticed that '…the slum is a ghetto…and the army of unemployed lives in this ghetto…Slavery still continues, only that it is different today. Now the individual doesn't have the right to survive, doesn't have the right to work, nor to be educated, or to be healthy and has to live under appalling living conditions' (interview conducted in late 1999).

The poor are becoming more organised and there is evidence of an increasing number of residents' associations in larger cities in Brazil. Usually these groups are action centred and focus on self-help projects. The Recycling Forum in São Paulo described earlier is an example of an emerging social movement. Recycling initiatives similar to the one described in *Pedra sobre Pedra* can contribute to diminishing irregular garbage disposal and improving sanitation and environmental awareness. This kind of recycling project has the potential to generate employment. Many people are needed in the collection, separation and commercialization of recyclables. The activity increases environmental consciousness because it promotes resource recovery and prevents littering and wasting. Finally, recycling initiatives have the capacity to empower the community and thereby strengthen participatory processes and build social cohesion. The creation of networks, such as the regional network for recycling, is an important step towards more efficient people's organisation and social inclusion. The experience has shown how locals are putting into practise Paulo Freire's popular education with strengthening critical awareness about the condition of poverty and in breaking *the culture of silence* (Freire 1970). Given the restricted budgetary circumstances, and the often inefficient municipal solid waste services, the local and state governments need to support and strengthen innovative ways to better manage domestic waste. The present case study indicates that recycling is a viable alternative to the conventional approach to urban solid waste management.

The externalities deriving from packaging, consumption and disposal, part of the true waste management cost, are usually not taken into consideration in the conventional cost-benefit calculations. Once we start to consider these the ineffectiveness of conventional waste management will be uncovered, providing an opportunity for preventive and inclusive waste management strategies to emerge. Recycling is still widely perceived as not being cost-effective enough. Considering the benefits mentioned above and giving them a deserved credit, recycling can indeed be a competitive and feasible option, tackling social and environmental problems at the same time, particularly under the described circumstances prevailing in countries like Brazil.

Cost recovery and revenue generation are aspects of economic sustainability that must be considered and require, according to Ali (2003), entrepreneurial tactics. While economies of scale can enable larger waste enterprises to reduce costs, community-based enterprises benefit from the social assets that come from collective initiatives, such as high participation rates in recycling schemes. These social assets translate into cheaper operations than those achieved by large-scale processes, even considering economies of scale.

The literature describes individual collection of recyclable materials, voluntary recycling activities, community recycling and professional small-scale recycling in Africa, Asia and Latin America (Chung and Poon 1999, 1998; Ferguson and Maurer 1996; Halla and Manjani 1999; Hernández et al. 1999; Rabinovich 1992). Most examples have shown that, to be successful, some form of financial support and infrastructure is needed. The local and state governments as well as other stakeholders have to start valuing the benefits deriving from these activities. Recycling improves environmental quality by recovering resources that otherwise would end up in the conventional waste stream or at irregular dumping sites. Recycling further promotes environmental education by creating awareness about resource flows and the limits of resource exploitation. Finally, it contributes to social sustainability by creating employment and by diminishing poverty.

To introduce community recycling, local leadership and a high degree of public participation is required. Without support from the local government, the business or non-governmental sector, these projects cannot flourish and their valuable social and environmental contribution is lost. Political will is a prerequisite to achieve successful recycling. Funding and infrastructure support are essential to facilitate the collection, separation, transportation and trade of recyclables. Local residents, industries, NGOs, and politicians are all stakeholders in the generation and disposal of waste, and together they have to bring about fair resolutions to the resulting social, economic and environmental problems.

The case of *Pedra sobre Pedra* provides evidence that it is certainly possible to improve the quality of life in squatter settlements, and that downstream impacts can be diminished through recycling. The *Local Agenda 21*, launched at the 1992

Earth Summit, advocates active citizenship. The engagement of citizens in local development issues and the establishment of partnerships and collaborations among the community, government and business sectors are vital in the change towards a more sustainable society. Barton (2000) has argued that the biggest hurdle for people is the paradigm shift away from competition between interests, towards the practice of co-operation. He refers to this process as '...creative policy-making that fulfils all the criteria [that is]...socially inclusive, economically viable, resource conserving and aesthetically pleasing' (Barton 2000, 8). Many countries in the South face similar scenarios to the one described in the case study, where a major hurdle for socially inclusive and ecologically sustainable development is the fact that 'creative policy-making' usually does not include the periphery, nor does it consider marginalised populations. And yet, under the *Agenda 21,* local participation of the main stakeholders in the decision-making process is crucial to achieving: '...policy consistency, shared ownership and commitment' (Barton 2000, 247).

New forms of governance are required to address and resolve the deterioration of the quality of life and the environmental conditions at the urban fringe. *Good governance* equals good administration, and it involves the democratization of politics. It concerns the relations between the government and civil society and it promotes citizen's participation and responsibility in the development process (Swilling and Hutt 2001). This means local participation in the decision-making process. It is through the eyes of local people that problems become transparent and more adequate and realistic solutions can surface. In many municipalities in Brazil governments have already began to respond to the issues of poverty and exclusion. Abers (1998) analyses *participatory budgeting,* practiced in Porto Alegre since 1989, where more than 14,000 people participate in the process of determining how the city has to invest to change their neighbourhoods. Local leaders and representatives take part in the elected municipal budget council that has deliberative power over the city's expenditures. *Participatory budgeting* has great potential to facilitate and secure effective participation in decision-making. It provides a formal structure that allows for participation and creative renewal. The model has been copied to several other municipalities in Brazil and abroad (Abers 1998). This governance form can bring policy formulation and decision-making closer to citizens themselves. When successfully applied, such strategies can reduce the risk of corruption and include the priorities of excluded groups into decision-making arenas. Hence there is a great potential to increase the overall responsiveness of local governments. However, as noted by Beall, '...decentralization does not on its own automatically lead to such outcomes' (2000, 853). Chapter 6 will discuss concrete examples on how to promote good governance by implementing inclusive waste management concepts.

Finally, the introduction of the geographic unit of water catchment areas in regional administration is another step that facilitates the increase of direct community participation in the definition of resource use. Watershed committees have now been introduced throughout most of Brazil. However, the communication and participation link between the local and the regional needs to be built in order to effectively express community concerns and to search for feasible, local solutions. Neighbourhood associations play a significant role in voicing local concerns, spearheading neighbourhood efforts and building communities with stronger social cohesion. They contribute to what is vital in this process of strengthening social capital in the community: a place to meet, an opportunity to discuss and disseminate and a more powerful voice to be heard in other political arenas.

Under sustainable communities, the focus is on closed resource loops, impact reduction, capacity building and human development in the community. Specific local, cultural, social and economic conditions necessarily have to come into play in the design of more sustainable communities and ultimately determine the scope and practice of each individual initiative. Community-based and co-operative recycling initiatives are new forms of collective actions tackling income generation and environmental health problems. The examples have shown how individuals and communities can become empowered and social capital can be constructed within the process of community-based recycling.

References

Abers, R. (1998), 'Learning democratic practice: Distributing government resources through popular participation in Porto Alegre, Brazil', in Douglas, M. and Friedmann, J. *Cities for Citizens* (Chichester: Wiley & Sons).

Ali, M. (2003), 'Community-based enterprises: constraints to scaling up and sustainability', in 'Report of the Workshop on Solid Waste Collection that Benefits the Urban Poor', *Collaborative Working Group on Solid Waste Management in Low- and Middle-income Countries*.

Barton, H. (2000a), 'Conflicting perceptions of neighbourhood', in Barton, H. (ed.), *Sustainable Communities* (London: Earthscan).

—— (2000b), 'Towards sustainable communities', in Barton, H. (ed.), *Sustainable Communities* (London: Earthscan).

Baud, I., Grafakos, S., Hordijk, M. and Post, J. (2001), 'Quality of life and alliances in solid waste management. Contributions to urban sustainable development', *Cities* 18(1), 3–12.

Beall, J. (2000), 'From the culture of poverty to inclusive cities: Re-framing urban policy and politics', *Journal of International Development* 12, 843–56.

Besen, R. (2006), 'Programa de coleta seletiva de Londrina – caminhos inovadores rumo à sustentabilidade', in Jacobi, P. (ed.), *Gestão Compartilhada dos Resíduos Sólidos no Brasil* (São Paulo: Anna Blume).

Brady, D. (2003), 'Rethinking the sociological measurements of poverty', *Social Forces* 81(3), 715–51.

Cardan, P. (1974), *Modern Capitalism and Revolution* (London: Solidarity).

Cazetta, N.H. (2005), 'Gestão de coleta seletiva com inclusão social e a situação dos catadores de materiais recicláveis de Santo André', Monograph (Graduate Thesis in Environmental Management, Santo André, Centro Universitário Fundação Santo André).

CEMPRE (2000), 'Compromisso Empresarial para Reciclagem, São Paulo', [website] http://www.cempre.org.br, accessed 15 June 2000.

Chaskin, R.J., Gocrge, R.M., Skyles, A. and Guiltinan, S. (2006), 'Measuring social capital: an Exploration in community – Research partnership', *Journal of Community Psychology* 34(4), 489–514.

Chung, S.-S. and Poon, C.-S. (1998), 'A comparison of waste management in Guangzou and Hong Kong', *Resources, Conservation and Recycling* 22, 203–216.

—— (1999), 'The attitudes of Guangzhou citizens on waste reduction and environmental issues', *Resources, Conservation and Recycling* 25, 35–59.

Durning, A.B. (1989), 'Action at the grassroots: fighting poverty and environmental decline', *Worldwatch Paper* 88 (Washington DC: Worldwatch Institute).

Ferguson, B. and Maurer, C. (1996), 'Urban management for environmental quality in South America', *Third World Planning Review* 18(2), 117–54.

Foster, M. and Mathie, A. (2003), 'Situating asset-based community development in the international development context', [website] http://www.stfx.ca/institutes/coady/about_publications_new_situating.html, accessed 24 March 2003.

Freire, P. (1970), *Pedagogy of the Oppressed* (New York: Continuum).

Halla, F. and Manjani, B. (1999), 'Innovative ways of solid waste management in Dar Es Salaam: Toward stakeholder partnerships', *Habitat International* 23(3), 351–61.

Hernández, O., Rawlins, B. and Schwarts, R. (1999), 'Voluntary recycling in Quito: factors associated with participation in a pilot programme', *Environment and Urbanization* 11(2), 145–59.

Hjorth, P. (2003), 'Knowledge development and management for urban poverty alleviation', *Habitat International* 27, 381–92.

Hornby Recycles, (2007), [website] http://www.hornbyisland.com/Recycle/index.html, accessed 19 August 2007.

Institute on Governance (2007), 'Learning tools', [website] http://www.iog.ca/boardgovernance/html/gov_wha.html, accessed 19 August 2007.

Iyer, A. (2001), 'Community experience in waste management. Case study report', *WASTE Advisers on Urban Environment and Development*, [website] http://www.waste.nl/, accessed 1 July 2005.

Kriesi, H., Koopmans, R., Duyvendak, J.W. and Giugni, M.G. (1995), *New Social Movements in Western Europe: A Comparative Analysis* (Minneapolis: University of Minnesota Press).

Legros, M. (2004), 'Against poverty: A common measure', *International Review of Administrative Sciences* 70(3), 439–53.

Luckin, D. and Sharp, L. (2004), 'Remaking Local Governance through Community Participation? The Case of the UK Community Waste Sector', *Urban Studies* 41(8), 1485–505.

—— (2005), 'Exploring the community waste sector: Are sustainable development and social capital useful concepts for project-level research?' *Community Development Journal* 40(1), 62–75.

—— (2006), 'The community waste sector and waste services in the UK: Current state and future prospects', *Resources, Conservation and Recycling* 47, 277–94.

Maser, C. (1997), *Sustainable Community Development: Principles and Concepts* (Delrey Beach: St. Lucia Press).

Medina, M. (2000), 'Scavenger Cooperatives in Asia and Latin America', *Resources, Conservation and Recycling* 3, 51–69.

Miraftab, F. (2004), 'Making Neo-liberal Governance: The Disempowering Work of Empowerment', *International Planning Studies* 9(4), 239–59.

Mitlin, D. (2003), 'Addressing urban poverty through strengthening assets', *Habitat International* 27, 393–406.

Muller, M.S., Iyer, A., Keita, M., Sacko, B. and Traore, D. (2002), 'Differing interpretations of community participation in waste management in Bamako and Bangalore: some methodological considerations', *Environment and Urbanization* 14(2), 241–57.

Painter, J. (1995), *Politics, Geography and Political Geography: A Critical Perspective* (London: Arnold).

Panelli, R. (2004), *Social Geographies* (London: Sage).

Programa Nacional Lixo e Cidadania, (n.d.), [website] http://www.rebidia.org.br/novida/reliase2.html, accessed 19 January 2006.

Rabinovich, J. (1992), 'Curitiba: towards sustainable urban development', *Environment and Urbanisation* 4(2), 62–73.

Robbins, C. and Rowe, J. (2002), 'Unresolved responsibilities: exploring local democratization and sustainable development through a community-based waste reduction initiatives', *Local Government Studies* 28(1), 37–58.

Romani, A.P. (2004), *O Poder Público Municipal e as Organizações de Catadores*, (IBAM/DUMA/Caixa Econômica: Rio de Janeiro).

Room, G.J. (1999), 'Social exclusion, solidarity and the challenge of globalization', *International Journal of Social Welfare* 8, 166–74.

Roseland, M. (1998), *Toward Sustainable Communities: Resources for Citizens and their Governments*, (Gabriola Island, BC: New Society Publishers).

Rowlands J. (1995), 'Empowerment examined', *Development in Practice, 1995* 5(2), 101–106.

Sandercock, L. (2002), 'Difference, fear and habitus: a political economy of urban fears', in Hillier, J. and Rooksby, E. (eds), *Habitus: A Sense of Place* (Aldershot: Ashgate).

SBPC (2002), 'Lixo é problema ambiental com agravantes sociais', in *Cidades. Sociedade Brasileira para o Progresso da Ciencia*, [website] http://www.comciencia.br/reportagens/cidades/cid10.htm, accessed 9 January 2006.

Sen, A. (1996), *Inequality Re-examined* (New York: Russell Sage Foundation).

Shriver, T.E., Cable, S., Norris, L. and Hastings, D.W. (2000), 'The role of collective identity in inhibiting mobilization: solidarity and suppression in Oak Ridge', *Sociological Spectrum* 20(1), 41–64.

Silver, H. (1994), 'Social exclusion and social solidarity: Three paradigms', *International Labour Review* 133, 531–78.

Simon, S. (2003), *Sweet and Sour. Life-worlds of Taipei Women Entrepreneurs* (Oxford: Rowman and Littlefield Publishers).

Swilling, M. and Hutt, D. (2001), 'Johannesburg, Afrique du Sud', in Onibokun, A.G., *La Gestion des Déchets Urbains, des Solutions pour l'Afrique* (Paris, Karthala, Ottawa: CRDI).

Timbo, F. (2003), *Participation, Negotiation and Poverty: Encountering the Power of Images: Designing Pro-poor Development Programmes* (Burlington: Ashgate).

UNFPA (2001), *The State of World Population 2001*, Chapter 3: Development Levels and Environmental Impact, [website] http://www.unfpa.org/swp/2001/english/ch03.html, accessed 1 July 2007.

Wagle, U. (2002), 'Rethinking poverty: definition and measurement', *International Social Science Journal* 54(1), 155–64.

Young, A., Russell, A. and Powers, J.R. (2004), 'The sense of belonging to a neighborhood: can it be measured and is it related to health and well being in older women?', *Social Science and Medicine* 59, 2627–37.

Chapter 5

Wasting Health[1]

Introduction: Occupational Health and Informal Recycling

In the previous chapters I discussed the scope of organized and informal recycling among the urban poor in countries with large socio-economic disparities. The extent of health problems related to waste is less well known. Waste management has to deal with the cumulative impacts, from air, water to soil contamination. Hamer (2003) informs about the effects of solid waste treatment and disposal on public health and environmental safety. Informal recyclers face serious human health predicament. Occupational health seems to be silently threatening this population during their daily working lives.

The wellbeing of recyclers is at risk primarily because of socio-economic exclusion and poor living conditions. There are also severe issues of unsafe working conditions. This chapter focuses on the health problems and the occupational risks related to handling recyclables. I will focus only on the most tangible health problems of informal recyclers as perceived by themselves. Informal recyclers have no support infrastructure to do the resource separation, compacting and commercialization. They work for daily subsistence and usually suffer lifelong social exclusion. The literature primarily talks about diverse occupational health and risks among the formal refuse workers in developing and developed countries. It also points towards possible risks at recycling facilities, particularly due to aerosol contamination from dirty recyclables as well as the risks of accidents at the work place (Lavoie and Guertin 2001; Medina 2000). Only very few studies tackle the health problems of informal recyclers separating in the street or at the waste dump. Hundreds of thousands of informal recyclers work for more than ten hours a day, seven days a week, sunshine or rain, throughout the year. Their livelihood issues are barely reflected in the academic literature. They are permanently facing risky situations while collecting, separating and transporting the material in improvised carts. They are stigmatized and sometimes even physically violated by the police and the public, and yet they are doing a precious service for humanity.

A case study will be presented to highlight the health issues of informal recyclers. Together with my Brazilian colleague Angela Baeder and three Masters students (Noé Humberto Cazetta, Ivan Corrêa and Vilson Rodrigues da Costa) from Centro

1 The case study was first presented at the *XXVII International Congress of the Latin American Studies Association* (LASA), in San Juan, Puerto Rico, 15–18 March 2006. The findings have been published in the International Journal for Environmental Health Research. Funding was provided by: *MSFHR-UVic Health Research Grant Preparation Program for New Investigators*.

Universitário Fundação Santo André, we conducted an in-depth socio-economic survey with informal recyclers in the city centre of Santo André. Our findings about the main health concerns are consistent with what has been described in the literature. Almost all workers report frequent body pain or soreness in the back, legs, shoulders, and arms. Injuries, particularly involving the hands, are quite frequent and the most common illnesses are related to catching a cold, flu and bronchitis. The risk of getting an infectious disease is high. One recycler had contracted Hepatitis-B.

Policy makers at all government levels need to address the pressing health concerns affecting large numbers of informal recyclers in Brazil and abroad. The recyclers working in associations or co-operatives also face risky working conditions. Because of their level of organization it is easier to reach the recycling co-ops and associations with policy measures and training to reduce occupational health risks. In some cases the recyclers in co-ops have already received training from the government and therefore have more information about health impacts and are more aware about their working situations.

This chapter will conclude with a discussion of the quest for participatory strategies in developing prescriptions for precaution and implementing measures that address these health problems. Recyclers, similar to other stakeholders, need to be involved in the design of waste management policies. Their daily experience of handling recyclables and dealing with the public can contribute to finding better ways of managing our waste. The wider community must be educated about the important environmental service the recyclers provide. Otherwise, the recyclers will continue to be marginalised and stigmatized.

Consumption, Waste and Environmental Health

Consumption and waste disposal rates are rising everywhere. In poor countries often only part of the household waste is regularly collected, and a fraction of it is disposed of safely. Deficient collection and inappropriate management of waste pose serious risks to human health and to the environment (Medina 2005). The collection and separation of recyclables also creates risks to the people involved in this activity (Cotton et al. 1999; Harpet 2003; Kennedy 2004; Sarkar 2003). Waste sorting poses microbiological, chemical, physical and ergonomic risks (Lavoie and Guertin 2001). An et al. (1999) argue that municipal waste workers are exposed to more occupational health and safety risks than workers in any other activities. This is true even in rich countries, as Poulsen et al. (1995) report for Denmark, where municipal solid waste workers faced 5.6 times the risk of occupational health injuries as the average total work force in the country, and 1.5 times the risk of occupational diseases.

Occupational health '…is a basic element [that] constitutes a social and a health dimension of the principle of sustainable development' (WHO 1995, 4). Occupational health risks relate to physical, chemical, and biological factors in the environment; they may also include economic and social determinants of occupational conditions, such as job security. According to the World Health Organization, environmental health comprises all encompassing aspects of human health, including quality of life, that are determined by physical, chemical, biological, social, and psychosocial factors in the environment (WHO 1995). Environmental health is further defined as the '... characteristics of health that result from the aggregate impact of both natural and [human]-made surroundings, including health effects of air pollution, water pollution, noise pollution, solid waste disposal, and housing; occupational disease and injuries; and those diseases related to unsanitary surroundings' (CDPH 2000).

In poor countries the informal sector is predominant and often determines the local economy. According to Santana and Loomis (2004), approximately half of the labour force in Brazil works without a formal job contract. Informal recycling in Brazil involves women and sometimes also children. As this is an unregulated activity, no occupational health services or supporting legislation are in place to address the workers' needs (Cazetta 2005). No matter what their level of organization, however, occupational health and safety are always major issues for recyclers.

Informal Recycling in Santo André

The city of Santo André is situated within the metropolitan Region of São Paulo. It is a city of 660,000 inhabitants in a territory of approximately 175 km² (IBGE 2004) (Figure 5.1). Almost 45 per cent of the area is densely populated. The remaining 55 per cent, located within the environmentally protected watershed area of Lake Billings, has a lower population density. The city is part of the metropolitan region of São Paulo. Living conditions in Santo André are similar to those in most urban agglomerations in Brazil.

The population in Santo André generates on average 0.85kg of solid waste per person (IPT/SEBRAE 2002). The city's current waste collection system includes door-to-door collection (general waste, humid residues, and differentiated collection of dry residues); voluntary hand-in stations (so-called PEVs, *Pontos de Entrega Voluntária*); and selective waste collection centres. Since 1997, Santo André has been engaged in the implementation of an integrated solid waste management plan. The plan focuses on the reduction, reuse, and recycling of solid waste and on integrating educational measures to maximize source separation and recovery. So far there is no specific programme to include autonomous, informal recyclers (Corrêa 2005). In 1998, the existing waste collection and recycling programme *Reciprocidade Agradável* was integrated into the agency SEMASA (Serviço Municipal de Saneamento Ambiental de Santo André) now in charge of the co-ordination of the city's waste management system.

During the same year, the municipal secretary for economic development and income generation (Secretaria Municipal de Desenvolvimento Econômico e Geração de *Renda*) encouraged the creation of the recycling co-operative, *Coopcicla*, as a work and income generating measure under the municipal programme *Programa de Geração de Trabalho e Renda*. Today, approximately 80 formerly unemployed and autonomous workers are affiliated with *Coopcicla*. The rise in household collection of recyclables by the city led to the creation of another co-operative, *Coop Cidade Limpa*, in 1999. This group of approximately 110 recyclers collects, separates, and commercializes recyclables from shantytowns that do not have a regular waste collection service. Officials estimate that at least 2,000 of Santo André's 64,000 unemployed work in recycling (Okabayashi et al. 2004). Nevertheless, the quantity of material recovered

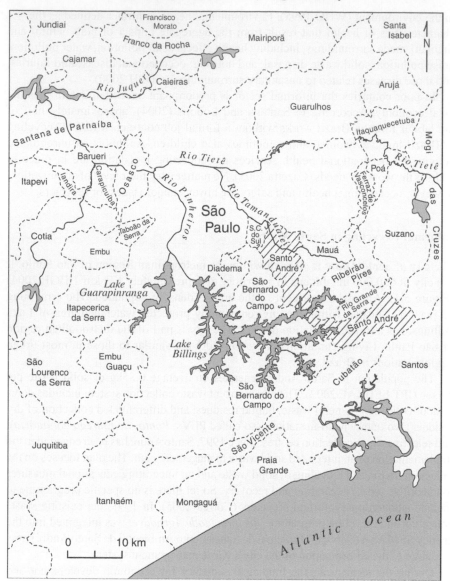

Figure 5.1   Localization of Santo André

Source: (IPT/SEBRAE 2002).

Cartography: Ole J. Heggen.

from the waste stream remains relatively small. Only 1.9 per cent is recovered, compared to the average 1.8 per cent for the state of São Paulo, which includes many cities that do not yet have a formal recycling programme (Cazetta 2005). The following figure shows the scope of selective collection in Brazil (Figure 5.2).

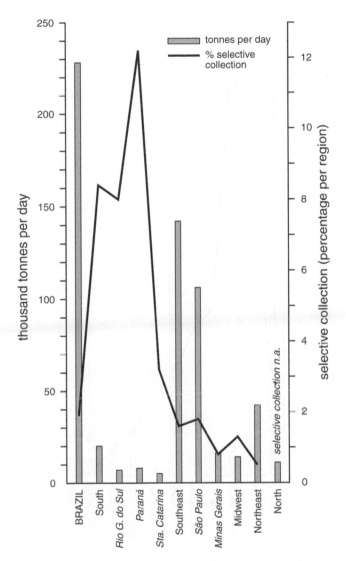

Figure 5.2   Selective collection of recyclables by region (in 10,000 tons/day)

Source: Cazetta, N.H. (2005), *Gestão de coleta seletiva com inclusão social e a situação dos catadores de materiais recicláveis de Santo André* [Graduate Thesis in Environmental Management]. Santo André (Brazil): Centro Universitário Fundação Santo André.

Cartography: Ole J. Heggen and Jutta Gutberlet.

On average, only 6 per cent of all households in Brazil are serviced by selective waste collection (Figure 5.3). In the state of São Paulo, only 5.4 per cent of the households receive this service. The official policy of Santo André is to provide every household in the future with selective waste collection (Cazetta 2005).

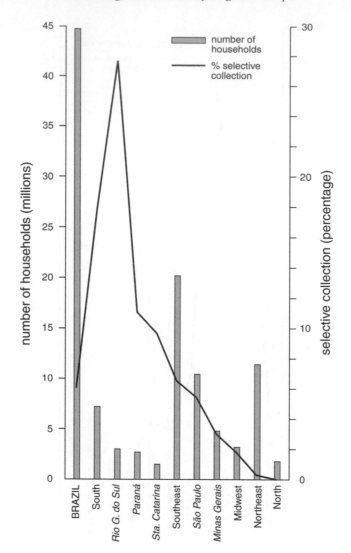

**Figure 5.3　Number of households participating in selective collection
(in millions)**

Source: Cazetta, N.H. (2005), *Gestão de coleta seletiva com inclusão social e a situação
dos catadores de materiais recicláveis de Santo André* [Graduate Thesis in Environmental
Management]. Santo André (Brazil): Centro Universitário Fundação Santo André.

Cartography: Ole J. Heggen.

Members from the co-operatives reveal that the sorting by householders is still
very poor, and that 30 per cent or more of the material collected for recycling is
contaminated and needs to be discarded.

Despite the increasing debate on solid waste management and environmental health, surprisingly little research has been done in developing countries on human health risks associated with informal domestic waste collection and handling. Few articles focus on the occupational health of informal recyclers in North America (Rendleman and Feldstein 1997) or in Brazil (Porto et al. 2004). In the literature on Brazil, most research discusses the health situation of formal municipal refuse workers only (Pereira 1983; Ilario 1989; Robazzi et al. 1994a, 1997b; Velloso 1995; Fundacentro 1999; Santos 2005). Research in other developing countries discusses similar topics within the formal waste management system (Poulsen et al. 1995; Lavoie and Guertin 2001; Harpet 2003). Literature on health-related waste issues tends to provide a general overview, rather than offer quantifiable data on health impacts that can be monitored over time or provide potential innovative solutions.

Our study focused on the recyclers' perceptions of health conditions and possible occupational risks, in order to be able to suggest solutions to problems concerning the livelihoods and health of recyclers. The study included additional questions, which provided the empirical data for this research on their perception about their physical health and occupational safety.

The survey was applied to the informal recyclers working near the recycling facility *Estação de Coleta Seletiva Bosque*, in the city centre. The study site was selected based on prior contacts with informal recyclers during capacity building activities, provided in 2002 by the city in partnership with university professors from FSA (Cazetta 2005). Land use and commercial activities in the city centre are diverse and their scale varies in size, with some industries, residential areas, and heavy pedestrian traffic around the train station. Therefore we limited our data collection area to a small radius in the immediate vicinity of the recycling facility, an area of approximately 2.3 km².

Structured interviews and participant observation were used for data collection (Gil 1991). The fieldwork was divided into three phases: (i) preliminary study involving initial contacts with the local recyclers, the *catadores*, with the research team visiting the area at different times of the day in order to identify the recyclers, the type of work they do, and the specific locations and land use characterizing the area; (ii) mapping of the information previously observed and logistic fine-tuning of the fieldwork; and (iii) observing participants and conducting an in-depth survey applying 36 standardized questions with a total of 47 interviewees. The recyclers were approached randomly while performing their work in the street. The interviews were conducted in the street, when the researchers met the *catadores* at their work. Each interview took between 20 and 30 minutes (see Illustration 5.1).

The results confirm a strong gendered labour division, with women more likely to work in the separation of the material and men in the collection and transport of recyclables. The youngest participant was 14 years old, and the oldest 64 years though most of the participants were between 21 and 60 years. We were interested in demographic information, quality of housing, work condition, transportation means, average daily income, formal education, general health and perception of work-related risks.

Illustration 5.1   Informal recycler in Santo André

Specific Work Related Health Risks of Informal Recyclers in Santo André Compared to Other Countries

With the following three health-related questions, we were able to cross-check the most relevant information: 1) Which health problems have you experienced after working with recycling? 2) What are your major persisting health problems? 3) Where do you feel frequent pain?

Eighteen interviewees responded that they hurt themselves at work quite often and nine mentioned that they had already experienced at least one accident while collecting recyclable material in the street. In most cases, they were hit or run over by a car. Further, seven respondents agreed with the statement that 'they had already experienced a problem with a driver, a pedestrian, or the police'.

When asked to compare their general state of health since they started to work as informal recyclers, most of the participants did not perceive any change. Only 15 of the interviewees referred to a new, specific health problem since working with recycling. Almost all *catadores* mentioned the common cold as the most frequent illness. Recurring problems with ulcers and high blood pressure, typical indicators for stress, were also reported quite often. Work accidents affecting the hands occur repeatedly. Two of the interviewees had already injured their hands (one had lost two fingers) before entering this type of activity, and two others had cut their hands while working with recyclables. More than half of the interviewees who continued having health problems after they had started in the recycling business live either in the street or in very precarious housing conditions (for example, slums or squatter settlements).

■ (gray) health problems since working in recycling

■ (black) persisting health problems since working in recycling

○ occurrence of frequent pain

number of informant	1	7	8	10	11	12	14	15	17	18	19	20	22	24	25	26	27	28	29	30	33	34	35	36	38	39	40	41	42	44	46	47
arrhythmia																													▨			
heart pain																	■															
high blood pressure																		■														
body pain						▨										○						○	■	○			○					
prostate						■																										
spine									○	○	○							○	○													○
back	○	○			○	○				○				○	○							○						◉				
chest																						○										
legs	○				○	○			○		○			○				○				○			○		○					○
feet																								○								
shoulders							○																									
arms	○						○															○										○
hands					■							■																				
loss of fingers																											▨					
finger cut													■																			
furuncle																				■												
ulcer																			■													
head																	▨	○														
headache																	■															
loss of consciousness																	■															
memory loss																		■														
labyrinthitis																			■													
throat					■																											
sinusitis							▨																									
bronchitis																					■						▨					
pneumonia																	▨															
flu								■					▨															▨				
convulsion													▨															▨				
weight loss																							▨									
hepatitis B																												■				
rheumatism																			■													
varicose veins																			■													
exhaustion		○																														

Figure 5.4   Detailed list of health problems

Credits: Data collected by Noé Humberto Cazetta, Ivan Corrêa and Vilson Rodrigues da Costa. Santo André (Brazil): Centro Universitário Fundação Santo André.

Cartography: Ole J. Heggen and Jutta Gutberlet.

More than half of the respondents (27) confirmed that they experience frequent pain. The most common kinds of pain were back pain (14), pain in the legs (nine), pain in the arms (four), and general body pain (four) (Figure 5.4). Refuse workers and recyclers often have to do heavy lifting, as well as pulling and pushing containers and carts on a daily basis. All of these activities present significant risk factors for lower back pain. Musculoskeletal injuries are often the result of continuously carrying heavy loads (An et al. 1999). Disorders involving the neck, shoulders, and arms were reported in different studies in the United States and in Denmark, as discussed in Poulsen et al. (1995). Towing these heavily loaded carts, uphill and downhill, through, often dense, traffic in the city centre puts particular strain on these body parts and can cause stress (Ackerman and Mirza 2001).

Another survey conducted in Florida found evidence of elevated rates of injuries such as lacerations, contusions, strains/sprains, and illness among waste collectors (Smith and Linton 2001). 75 per cent of the municipal collectors interviewed reported having been injured over a 12-month period (Smith and Linton 2001). The authors further mention that health incidents and accidents for this category are usually underestimated, because often the person who is injured or ill leaves the workforce and has therefore not been captured in the assessment.

Occupational Health and Safety of Informal Recyclers

Accidents among refuse workers are frequent and they are widely discussed in the literature for European countries and the United States. Poulsen et al. (1995) indicate a significantly higher level of occupational accidents among Danish refuse workers (95 per 1000 employees) than among the total work force in the country (17 per 1000 employees) (1995, 5). In Brazil, Robazzi et al. (1997) recorded an extremely high frequency of accidents. This study reveals that the major cause of accidents among municipal refuse workers was related to improper garbage wrapping. The body parts most affected are the legs, followed by the arms. The most frequent medical diagnoses are wounds, contusion and cutting injuries, and excoriations.

A study conducted among municipal workers in Rio de Janeiro highlights the fact that 80 per cent of all workers had suffered an accident during their employment in waste collection; and of these, 58 per cent had to be suspended from work during some time (Vellos et al. 1997). Recyclers who work at dumpsites are exposed to extreme conditions and suffer accidents more than other recyclers. Porto et al. (2004) conducted a study on health conditions of informal recyclers at the largest landfill in Rio de Janeiro. The nature of the working environment generates specific kinds of risks and accidents. The majority (71.7 per cent) of the informal recyclers at this site had already suffered an accident. Of the 267 accidents referred to in the study, the majority were related to cuts from glass (37 per cent), followed by perforations due to other materials (19 per cent), and falls (15 per cent). In addition, accidents involving informal recyclers and the landfill operator's machinery are frequent (Porto et al. 2004). The film *Estamira* (2006 production) shows people sorting recyclables at the waste dump *Gramacho* in the outskirts of Rio de Janeiro, constantly threatened by unloading trucks and bulldozers paving the material. Interviews with members from the recycling co-operative in Rio de Janeiro conducted during the second Latin American Congress of Recyclers in 2006 mentioned the frequent accidents involving these machines, confirming the grim truth of these images.

Collecting and transporting recyclables demands a lot of human energy, as most of the respondents push their carts manually. In order to make a living, recyclers have to work long hours. In our study the typical workday was eight hours or longer for 31 of the respondents. Most work six or seven days per week and many cannot afford to take weekends off; they need the daily income to stay alive and to feed their families. Almost all respondents confirmed that besides collecting they also separate the recyclables, either in the street, at home, or at the middlemen's premises.

The size of the self-made carts varies largely. 18 of the respondents generally do only one round trip per day, 20 of them do two trips a day, and the rest do up to five trips. The work effort is significant. The amount carried per day can vary from 30 to 1000 kg. 21 of the interviewees transport up to 150 kg/day, 19 transport between 200 and 500 kg/day, and four more than 500 kg/day.

Every second interviewee explained that they also collect construction waste – for example wood, bricks and metal. Irregular disposal of inert waste is a visible problem in the periphery, especially in the metropolitan region of São Paulo. Waste is often dumped along the streets and rivers, where it becomes a traffic risk and an environmental hazard. In Diadema, the city has recently engaged the recyclers in an initiative to collect construction waste and to transform the material into new building resources, such as bricks or gravel for road construction. Despite the high volume of discard, there are few programmes that aim to recover these resources.

Livelihood Issues and Wider Health Impacts

Human health in a broader sense is also influenced by work environment and housing conditions, access to safe public infrastructure and services at home and during work play a crucial role. Adequate housing for the poor is one of the largest bottlenecks in most large cities in Brazil. In the metropolitan region, the deficit in low-cost housing is visible, and squatting has become a necessity for many. Consequently, many people do not have access to clean water and sewage collection, as already discussed in Chapters 2 and 3. In Santo André, the situation is particularly serious due to growing poverty rates caused by the recent de-industrialization and out-migration of industrial establishments and resulting high unemployment. In 2000, circa 64,000 people or 20.4 per cent of the economic active population, was unemployed. For women the rate was significantly higher (24.7 per cent) than for men (17.2 per cent) (Okabayashi 2004).

More than 15 per cent of the population of Santo André, or approximately 97,000 people, currently live under substandard housing conditions (Carneiro 2001). There are a total of 132 squatter settlements in Santo André, of which only 13 have recently been upgraded with basic infrastructure, amenities, and standard services; 37 have received partial improvements; and 82 remain with problematic living conditions. When low- or no-income groups do not own property and cannot afford to rent, they have no other choice than to squat, live in shared accommodation, homeless asylums, or live in the street (Carneiro 2001).

Our survey reveals that many interviewees were living under precarious housing conditions (Figure 5.5). The low and irregular income from informal recycling is usually insufficient to pay for housing and therefore squatting is one of the few

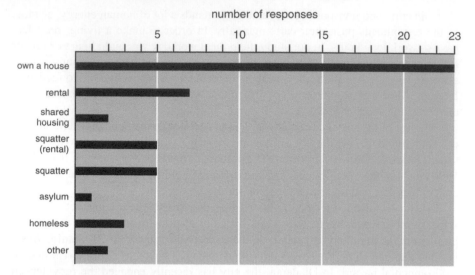

Figure 5.5   Housing situation

Credits: Data collected by Noé Humberto Cazetta, Ivan Corrêa and Vilson Rodrigues da Costa. Santo André (Brazil): Centro Universitário Fundação Santo André.

Cartography: Ole J. Heggen and Jutta Gutberlet.

options. However, squatting is inherently uncertain. The permanent threat of removal adds another stress factor to the already insecure livelihoods of these individuals, making them more vulnerable.

Most of the recyclers collect and separate the material. Very few (four) mentioned that others, most probably other family members, do the sorting. Most respondents (21), said that they separate the material in the street, and others (eight) separate at the dumpsite or at the middlemen's premises. Eleven of the interviewees sort the material at home. This means bringing disease vectors and risks home, which also exposes other family members to these risks. The recyclables usually are not clean, and having them close to their living space attracts insects, rodents, fungi, bacteria, and other transmitters of disease into the home. Fly infestation at home can be a health threat. Boadi and Kuitunen (2005) have reported a strong relation between the presence of flies in the household and the incidence of childhood diarrhoea.

Workers handling mixed waste, compost, and recycling products face a number of health problems (Poulsen et al. 1995). Several authors highlight the elevated risks of respiratory, dermal, and gastrointestinal problems among municipal waste workers, due to the exposure of dust, microorganisms, and microbial toxins at their workplace (Rogers et al. 2003; An et al. 1999; Poulsen et al. 1995). Mould and fungus spores, abundant on discarded food packaging, are well-recognized biological sources for respiratory diseases, mucous membrane irritation, and allergies in general (Poulsen et al. 1995). Dust and mould spores have the capacity to increase already existing allergic reactions. Waste exudates also contain Gram Negative bacteria, which in their outer membrane contain *endotoxins* (cell-associated bacterial toxins). *Endotoxins* are released when the bacteria are broken down or die, causing mucous membrane irritation. Often the workers are also exposed to chemical pollutants from containers contaminated with chemical substances, particularly when recyclers collect materials directly from industries (Vellos et al. 1997). The recyclers report that the material collected is usually quite dirty, that they don't wear gloves and cannot easily access washrooms when needed. This hazardous situation of discomfort causes health risks (see Illustration 5.2). The co-operative *Coopcicla* is one of the exceptions where the recyclers wear gloves and mouth protection. It has been reported though that this is not always the case.

Lavoie and Guertin (2001) found *organic dust toxic syndrome* to be a frequent health complaint at recycling plants, where contamination with bioaerosols can be very high. Bioaerosols are extremely small living organisms or fragments of living things suspended in the air. Dust mites, moulds, fungi, spores, pollen, bacteria, viruses, amoebas, fragments of plant materials are all examples of bioaerosols. Moulds and fungi that grow on unwashed food containers are major contaminants that can affect human health. Similarly, An et al. (1999) mention particular health problems that arise from the exposure of waste collectors to Gram Negative bacteria and fungal spores.

Storage and handling of recyclables are major sources of bacteria in the atmosphere. The concentration of microorganisms seems to fluctuate with the season, being greater during higher temperature and humidity levels in the summer.

Illustration 5.2 Minimization of occupational health risks in resource separation, *Coopcicla* Santo André

In addition, working in the street exposes waste collectors often continuously to automobile and truck exhaust gases that can also cause respiratory disease (Poulsen et al. 1995).

Our survey also inquired about the availability of regular meals while performing the job. All respondents indicated having access to at least one meal per day, with the exception of two recyclers who declined the answer to this set of questions. Both of them were homeless and lived in the streets. Three respondents, who were also either homeless or squatters, had only one meal per day. One of them had only breakfast as a guaranteed meal, whereas the other two had only lunch. Most of the respondents (26) however, had lunch and dinner, and 17 had breakfast and one hot meal during the day. The survey did not query the nutritious value of these meals. It is not rare to see recyclers consume food out of the garbage, exposing themselves to serious health risks. A situation of extreme humiliation and lack of human dignity.

It is very common among the recyclers to consume alcohol as a food substitute. Alcohol, a defence against the humiliation and stigma they frequently experience in the street, '...*works almost like a psychological mask, an armour to better face the other in the street, because it is not easy to be socially excluded in their life and in their work*' (interview with *Fundacentro* health agent, 7 April 2005). According to this health agent, alcohol provides 'empty calories' without adding necessary proteins. She further underlines that alcohol extracts crucial vitamins from the metabolism and over time the person suffers from vitamin B deficiency. Alcoholism is a major problem among this population; it consumes their scarce financial resources and it affects their health, but to the recyclers, it seems to diminish present suffering.

Medina has researched diverse aspects of recycling and waste management in Asia and Latin America, particularly in Mexico (1998, 2000, 2001, 2005). However, his work does not specifically focus on occupational health hazards. He underlines the pressing need to study the health effects and risks from collection and separation of recyclables (2001). Constant heavy work has its effect on the wellbeing of the worker. It is one of the factors contributing to the widespread spine and back problems of recyclers, and it can be a cause of general pain in the body, mentioned so often by most of the recyclers. Better collection and separation methods and infrastructure could alleviate some of the current ergonomic problems. Appropriate technology could be used to develop more efficient vehicles for material collection. Innumerable conversations with informal recyclers in Brazil confirm the fact that most recyclers do not object to using pushcarts, which allows access to places where cars and trucks cannot go. The recyclers also affirmed that owning the means of production provides them with autonomy. However, it seems urgent to improve the efficiency of the carts in order to make the work safer and less effort intensive.

Contracting infectious diseases is a real threat to refuse workers and particularly to informal recyclers. Rendleman and Feldstein (1997) underline the high risk of contracting Hepatitis B among solid waste workers in North America. In our study, we met one person who had contracted Hepatitis B during his work with recycling. Another widespread risk is the contraction of Leptospirosis, a bacterial disease, transmitted through the urine of infected rodents. Recyclers who handle waste, which frequently is the niche of rats infected by bacteria of the genus *Leptospira*, face this additional threat. The situation described for Accra, Ghana, reveals a variety of health-related symptoms associated with waste storage in households and waste burning, as well as environmental pollution and the spread of infectious diseases due to littering (Boadi and Kuitunen 2005).

Outlook: Recovering Citizenship and Health

Handling and processing recyclable materials exposes the workers to unhealthy and sometimes dangerous conditions. Despite the visibility of informal recyclers in the streets, their working conditions are hardly recognized, and they generally remain without a voice. The potential occupational health risks involved in this activity are also mostly unknown to the workers themselves. Significant improvements involve extended capacity building and education, the provision of adequate recycling facilities and the organization of the activity within the official waste management scheme of cities.

This survey of informal recyclers in Santo André reveals first insights to the substantial occupational risks these workers face in terms of injury and disease. The findings are consistent with the other literature on health issues involving the recycling sector. Most academic work discusses only the health predicaments of formalized employees at landfills and recycling depots, and little has been documented about the health situation of the many informal recyclers in developing countries. The aim of this chapter was to provide baseline data on typical occupational risks and injuries within this sector.

Occupational health risks are usually classified into the following five categories: (i) mechanical (cuts, blunt trauma, fractures, lacerations, traffic accidents); (ii) ergonometric (musculoskeletal illness resulting from moving heavy weights); (iii) chemical (dermatitis and respiratory disease due to exposure to toxic chemical substances); (iv) biological (infections due to contact with biological pathogens); and (v) social (as a consequence of poverty and social exclusion, stressful working conditions, and manifested negative perceptions against them by the public).

The recyclers' perceptions of their health are consistent with the findings in the wider literature on specific types and the frequency of injuries and disease that affect the refuse collectors and the formal recyclers elsewhere, particularly with the results from studies conducted in developing countries. Traffic accidents, which fall under the category mechanical health risks, were significant among the sampled population of informal recyclers. Nine of the surveyed recyclers had already suffered an accident while collecting or transporting materials in the street. Traffic congestions, dangerous conditions due to road obstructions, and on-road barriers add to the risks all participants face in the street. Recyclers with carts are among the most vulnerable participants in traffic. Visible clothing and reflectors on the pushcart as well as educational campaigns for a more humane traffic behaviour can prevent accidents and contribute to more safety for everybody involved in traffic.

Most of the recyclers hurt themselves frequently during their work. Cuts and fractures as a result of working with recyclables were very common and are most probably related to unsafe disposal of the recyclable material, which affects the recyclers during collection and separation. Source separation needs to be more efficient and safe. The public needs to collaborate with providing the materials clean and with disposing them safely. This means extensive and repeated environmental education campaigns at the household level and in schools.

The most frequent ergonometric health implication found in our base line study was body pain (in particular back pain), as a result of carrying heavy loads, constant bending, long working days and inadequate working conditions at the separation centres. Informal recyclers hardly have time for activities that aim at personal development. Ergonomic issues can be addressed. For example, the provision of battery-driven handcarts could reduce the weight to be carried and could reduce or prevent body pain and strains. Adequate and well-placed sorting infrastructure is crucial, as recyclers often spend many hours separating the materials. The setting needs to take into consideration specific gender-related needs. It is mostly women who are involved in the separation. Health impacts vary because of physical differences between women and men. Child bearing and menstruation, for example, make for physical differentiations and underline specific necessities. Women have also different needs due to their double function as care givers, particularly if they are single mothers. In many cases, women do not have a proper place to leave their children, which means that children are either locked inside the house (often for many hours), left alone in the street, or accompany the parents during their work. The recyclers spend most of their day working in the street. They depend on public infrastructure or favours to use the sanitary facilities. Sufficient and widely distributed community centres or recycling centres can provide the facilities to satisfy these basic needs, adding to the quality of life of the recyclers. *Fundacentro*, the federal research centre for occupational health in São Paulo, underlined the need to inform and educate the recyclers about personal hygiene and occupational safety (interview conducted on 7 April 2005).

Recyclers and refuse collectors are often exposed to extreme weather patterns (cold, hot, dry and wet), a contributing factor to catch colds, virus infections, sinusitis, or pneumonia. High levels of air pollutants cause additional health impacts, often exacerbating respiratory diseases, allergies, and skin problems. Sanitary and ergonometric conditions are controllable at recycling centres, and precautionary and protective measures can be implemented. Robazzi et al. show that with specific preventive training these risks of accidents can be diminished (1997). Access to public health care and provision of social benefits are also important needs and addressing these can help improve the health of informal recyclers. Regulated working hours is a factor that diminishes risks related to overwork. Formalizing groups such as co-operatives and recycling associations offers the possibility of containing these threats.

Dealing with contaminated material in closed environments puts informal recyclers particularly at risk to chemical contamination and biological infection. Kennedy et al. (2004) found that the most important point sources are mouldy containers. In most of the times the recyclers reported that the material is not clean when they collect it. Sometimes the recyclables are even mixed with toilet paper and rotten food waste. This could be changed with implementing an effective door-to-door educational outreach programme. Furthermore, proper ambient temperature and ventilation at the work place and the use of gloves and mouth protection can improve these circumstances. Providing tools alone is not sufficient they have to come together with education. Many recyclers have frequently expressed how uncomfortable working was when using gloves and mouth protection. Nevertheless, particularly indoors, respiratory health is influenced by exposures to airborne and surface mould. The recyclers need serious environmental and health education.

Attention also needs to be given to social health implications. The public often perceives the recyclers as a nuisance and react aggressively when they encounter recyclers with their pushcart in the streets. Powerlessness, vulnerability and low self-esteem are often the conditions that shape the experience of informal recyclers. They are frequently bullied or treated with prejudice by the public. They are widely stigmatized, and are socially and economically excluded by society. Sarkar's (2003) vulnerability study of rag pickers in Delhi, India, focused on the socio-economic and occupational health of these workers. They highlight their catastrophic health conditions as a consequence of poverty and low social status. In our study one of the interviewed recyclers reported, '*more respect would be the most needed change to improve work*'. A campaign is necessary to educate the public about recycling and the livelihoods of recyclers. This has the potential to expand partnerships between the recyclers and homes, as already happens (Lavoie and Guertin 2001). Kennedy et al. (2004) offer a number of suggestions for reducing health impacts, namely having consumers wash bottles prior to returning to remove fungal contamination (which can be done with minimal waste of water) and more stringent regulations and control for breaking glass or other containers with mould content.

Health concerns of informal recyclers receive little media or public attention, and the needs and health matters of informal recyclers are hardly ever taken into consideration in waste management policies. This sector does not have a lobby or pressure group to demand investigation into their specific health issues, and their cases remain unvoiced. Recyclers themselves most often are powerless and cannot afford to seek improved and healthier working conditions. Recyclers face so many pressing problems merely trying to survive, and their attention is usually concentrated on bringing home as many recyclables as possible to make it through the day.

Poverty is still a prevailing condition. Most informal recyclers make less than a minimum salary. The Brazilian recyclers movement estimates that autonomous informal recyclers in Brazil earned on average between Reais $60.00 and 100.00 per month in 2006, (Movimento Nacional dos Catadores de Materiais Recicláveis 2006). The value fluctuates significantly according to the material collected and the regional location, whereby large cities in the South and Southeast pay most. The average monthly income of the recyclers interviewed in this study was Reais $524 (approximately US$240). In Santo André, our study site, approximately 2,000 people live primarily from informal recycling. This reflects a situation similar to that in other urban centres in Brazil, where on average 0.3 per cent of the total population is involved in this activity. Providing opportunities for equitable and fair pay will help address the problems of inadequate housing, malnourishment and crime.

The risks and health issues examined here through the perspective of the recyclers themselves provide evidence of a clear lack of basic citizenship rights. These people are facing inadequate working conditions, severe health risks and are often stigmatized and humiliated by the rest of society, causing additional stress to their life. A first step towards recovering their citizenship needs to embrace the recognition of their role in improving environmental health, by paying for the service of resource recovery. Governments, and members of the public who are committed to changing the prevailing paradigm of wasting resources and ultimately inflicting changes on the climate, need to promote sustainable waste management strategies. One such strategy is the inclusion of the informal recycling sector as partner in waste management. With specific training on safer collection and separation practices, and by providing better working environments and improved source separation, the serious health issues can be tackled. Recognizing and formalizing the sector without infringing on the recyclers' autonomy is a proactive approach that will improve the catastrophic health conditions under which these people work.

While the collection of recyclables involves more men, there are more women participating in the separation of the material. Women are also a visible majority in recycling co-operatives and associations. The specific needs of women are still widely unknown. More research is necessary to better understand the needs of all recyclers in terms of health and safety, in order to influence legislation and public policies for better safety regulations.

The refuse workers' federation *Federação de asseio* has recently started to discuss the elaboration of specific norms for recycling workers. It needs to be underlined that refuse workers and *catadores* live and work in different realities. Furthermore, the *catadores* are not organized and don't yet have a union. The Brazilian norm for outdoor work, the NR21, is one of the legislation pieces under consideration (SINTRACOM-BA 2007). It was primarily elaborated for construction workers, and needs to be adjusted to the conditions of the highly mobile recyclers in the street. The norm regulates outdoor activities, and instructs about health-protecting practices to address the adverse health effects from weather conditions such as excessive heat, cold, humidity or wind. Informal selective waste collection programmes, public policies, and workers legislation need to address and guarantee payment for insalubrious work conditions, working under occupational risks, secure assistance in case of accidents and provide other work related benefits for this category. National safety codes for municipal waste workers need to be in place in order to guarantee minimum health standards.

A social paradigm shift needs to happen in order to qualify this work as an *environmental service*. Because resource recovery reduces emission of greenhouse gases, it is also a form of combating climate change. It is a way of recovering material that would otherwise be dumped, landfilled, or incinerated, possibly ending up in our watercourses and over time contributing to the generation of CO_2. Collecting these resources and redirecting materials into the recycling streams also prevents the extraction of virgin resources and thus indirectly contributes to resource and energy conservation. Recyclers are environmental agents because they conduct door-to-door collection; they educate and communicate with the local community. In order to be recognized as such, they need to be trained and made identifiable, for example by uniforms and ID cards. The research evidences that formalizing the appearance of the workers would contribute to a better relationship with the community, and would build their self-esteem. Although it might seem that such a measure would infringe on the independence of the recyclers the interviews reiterate the benefits of wearing a uniform. It would be a way of making recyclers feel that they are part of the public service. Further research is needed to confirm how formalizing the activity, for example, through recycling co-ops or government programmes, improves occupational health and diminishes risks.

In recent years, the city of Santo André has become a showcase for innovative governance, with community participation in budgeting decisions as well as other innovative practices and programmes in the areas of income generation, housing, and education. However, the local government is still not committed to provide the necessary support to the informal recycling sector and to widely include it in these new programmes. With increasing costs of conventional waste treatment, most governments are now in favour of recycling; however, seldom do they consider the informal recycling sector. There is hope that the prospect of generating income for the urban poor will provide an incentive for adopting innovative and inclusive community recycling management practices.

References

Ackerman, F. and Mirza, S. (2001), 'Waste in the Inner City: Asset or Assault?', *Local Environment* 6(2), 113–20.

An, H., Englehardt, J., Fleming, L. and Bean, J. (1999), 'Occupational health and safety amongst municipal solid waste workers in Florida', *Waste Management and Research* 17(5), 369–73.

Bernstein, J. (1993), 'Alternative approaches to pollution control and waste management: regulatory and economics instruments', *Urban Management Discussion Paper No. 3* (Washington DC: The World Bank).

Boadi, R.O. and Kuitunen, M. (2005), 'Environmental and health impacts of household solid waste handling and disposal practices in third world cities: The case of the Accra Metropolitan area, Ghana', *Journal of International Health* 68(4), 22–36.

Carneiro, C.B.L. (2001), 'Programa Integrado de Inclusão Social (PIIS), Santo André', in Farah, M.F.S. and Barboza, H.B., *20 Experiências de Gestão Pública e Cidadania*. São Paulo: Programa Gestão Pública e Cidadania, [website] http://www.apsp.org.br/Boletins/momentoAPSP/momento1/pag345.htm, accessed 20 July 2006.

Cazetta, N.H. (2005), *Gestão de Coleta Seletiva com Inclusão Social e a Situação dos Catadores de Materiais Recicláveis de Santo André* (Santo André Brazil: Centro Universitário Fundação Santo André).

CDPH (Connecticut Department of Public Health), Policy, 'Planning and Analysis' (2000), *Looking Toward 2000 State Health Assessment, 2000 – Appendix O Glossary*, [website] http://www.dph.state.ct.us/OPPE/SHA1999/shacontents.htm, accessed 20 July 2006.

Corrêa, I. (2005), *Os Catadores de Recicláveis de Santo André* (Santo André, Brazil: Centro Universitário Fundação Santo André).

Cotton, A., Marielle, S. and Mansoor, A. (1999), 'The challenges ahead – solid waste management in the next millennium', *Waterlines* 17(3), 2–5.

Fundacentro (1999), *Análise dos Acidentes de Trabalho e Doenças Profissionais dos Trabalhadores das Empresas Prestadoras de Serviços de Limpeza Pública da Cidade de São Paulo no Período de 1990–1994* (São Paulo: FUNDACENTRO).

Gil, A.C. (1991), *Métodos e Técnicas de Pesquisa Social* (São Paulo: Atlas).

Hamer, G. (2003), 'Solid waste treatment and disposal: effects on public health and environmental safety', *Biotechnology Advances* 22, 71–9.

Harpet, C. (2003), 'From garbage dumps anthropology to an interdisciplinary research on health risk exposure', *Natures Sciences Societes* 11(4), 361–70.

Hoornweg, D., Thomas, L. and Varma, K. (1999), 'What a waste: solid waste management in Asia', *Urban Development Sector Unit – East Asia and Pacific Region*, The World Bank, [website] http://worldbank.org/urban/publicat/ whatawaste.pdf, accessed 20 July 2006.

IBGE (Instituto Brasileiro de Geografia e Estatística) (2004), 'Ministério do Planejamento. Indicadores de desenvolvimento sustentável', [website] http:// www.IBGE.gov.br, accessed 20 July 2006.

Ilario, E. (1989), 'Estudo de morbidade em coletores de lixo de um grande centro urbano', *Revista Brasileira de Saúde Occupacional* 68(17), 7–13.

IPT/SEBRAE (2002), *Cooperativa de Catadores de Materiais Recicláveis: Guia Para Implantação*. Roberto Domenico Lajolo. 2002, 112. São Paulo: Instituto de Pesquisas Tecnológicas/SEBRAE, and CD (Publication IPT; 2952).

Kennedy, S.M., Copes, R., Bartlett, K.H. and Brauer, M. (2004), 'Point-of-sale glass bottle recycling: indoor airborne exposures and symptoms among employees', *Occupational Environmental Medicine* 61, 628–35.

Lavoie, J. and Guertin, S. (2001), 'Evaluation of Health and Safety Risks in Municipal Solid Waste Recycling Plants', *Journal of the Air and Waste Management Association* 51, 352–60.

Medina, M. (1998), 'Border scavenging: a case study of aluminum recycling in Laredo, TX and Nuevo Laredo, Mexico', *Resources, Conservation and Recycling* 23, 107–126.

—— (2000), 'Scavenger cooperatives in Asia and Latin America', *Resources, Conservation and Recycling* 31, 51–69.

—— (2001), 'Scavenging in America: Back to the Future?', *Resources, Conservation and Recycling* 31, 229–40.

—— (2005), 'Serving the Unserved: Informal Refuse Collection in Mexican Cities', *Waste Management Research* 23, 390–97.

Movimento Nacional dos Catadores de Materiais Recicláveis (2006), 'Proposta apresentada em 24 de janeiro de 2006', São Paulo (Resumo Executivo) (*unpublished report*), 6.

Okabayashi, A.M.H., de Paula, R.T.A. and Gimenes, S.P. (2004), 'Perfil socioeconômico da população feminina de Santo André em 2000' (Prefeitura de Santo André, Secretaria de Orçamento e Planejamento Participativo) (*report*).

Pereira, E. da S. (1983), 'Condições de saúde ocupacional dos lixeiros de São Paulo', *Revista Brasileira de Saúde Ocupacional* 42:11, 30–35.

Porto, M.F. de S., Juncá, D.C. de M., Gonçalves, R. de S. and Filhote, M.I. de F. (2004), 'Lixo trabalho e saúde: um estudo de caso com catadores em um aterro metropolitano no Rio de Janeiro, Brasil', *Caderno de Saúde Pública* 20:6, 1503–514.

Poulsen, O.M., Breum, N.O., Ebbehoj, N., Hansen, A.M., Ivens, U.I., van Lelieveld, D., Malmros, P., Matthiasen, L., Nielsen, B.H. and Nielsen, E.M. (1995), 'Collection of domestic waste. Review of occupational health problems and their possible causes', *The Science of the Total Environment* 170, 1–19.

Rendleman, N. and Feldstein, A. (1997), 'Occupational injuries among urban recyclers', *Journal of Occupational and Environmental Medicine* 39(7), 672–5.

Robazzi, M.L. do C., Moriya, T.M., Favero, M., Lavrador, M.A.S. and Luis, M.A.V. (1997), 'Garbage collectors: occupational accidents and coefficients of frequency and severity per accident', *Annals of Agricultural and Environmental Medicine* 4, 91–6.

Robazzi, M.L. do C., Moriya, T.M. and Pessuto, J. (1994), 'The trash collection service: occupational risks versus damages to health', *Revista da Escola de Enfermagem USP* 28(2), 177–90.

Rogers, J., Englehardt, J., An, H. and Fleming, L. (2003), 'Solid waste collection health and safety risks – survey of municipal solid waste collectors', *Journal of Solid Waste Technology and Management* 28(3), 154–60.

Santana, V.S. and Loomis, D. (2004), 'Informal Jobs and Non-fatal Occupational Injuries', *Annals of Occupational Hygiene* 48(2), 147–57.

Santos, T.L.F. (2005), *Relato de Experiência Programa Trabalhadores de Rua: Estudos e Intervenção* (São Paulo: FUNDACENTRO).

Sarkar, P. (2003), 'Solid Waste Management in Delhi – A Social Vulnerability Study', in Bunch, M.J., Suresh, V.M. and Kumaran, T.V. (eds), *Proceedings of the Third International Conference on Environmental Health*, Chennai, India.

SINTRACOM-BA 'Sindicato dos Trabalhadores na Indústria da Construção e da Madeira' (2007), Homologação – Instruções, [website] http://www.sintracom.org.br/homologacoes.html#01, accessed 14 September 2007.

Smith, D.N. and Linton, D. (2001), 'Health and safety issues in post-consumer aerosol container recycling', *Resources, Conservation and Recycling* 31, 253–63.

Vellos, M.F., dos Santos, E.M. and dos Anjos, L.A. (1997), 'Processo de trabalho e acidentes de trabalho em coletores de lixo domicilar na cidade do Rio de Janeiro, Brasil', *Cadernos de Saúde Pública* 13(4), 693–700.

Velloso, M.P. (1995), *Processo de Trabalho da Coleta de Lixo Domiciliar: Percepção e Vivência dos Trabalhadores* (São Paulo: Fundação Oswaldo Cruz, Escola Nacional de Saúde Pública).

WHO (World Health Organization) (1995), 'Global strategy on occupational health for all: The way to health at work', Recommendation of the second meeting of the WHO Collaborating Centres in Occupational Health, 11–14 October 1994, Beijing, China, 4, [website] http://www.who.int/occupational_health/publications/glob strategy/en/index3.html, accessed 20 July 2006.

Chapter 6

Co-management in Resource Recovery[1]

Introduction: Towards a New Perception of Waste Management

Innovative proposals are emerging in solid waste management all over the world. In this chapter I will discuss some of these alternative, inclusive arrangements, and show how they address pressing environmental and economic problems deriving from the generation of waste. Waste management is not solely about technical solutions to waste disposal, but is more so about social and environmental concerns. The rationale for such new forms of waste management is straightforward and logical, since they offer concrete solutions for social predicaments related to high unemployment, particularly in the developing world; they also help solve environmental concerns linked to solid waste disposal. Nevertheless, the application of policies to implement such new ways to manage waste is difficult. Existing paradigms and power structures must be changed. Participatory structures are required; however, these are sometimes interpreted merely as challenges to the status quo, and often governments and communities do not recognize their enormous potential as modes of more sustainable waste management. These novel proposals can affect existing political and economic interests, and are therefore often seen with scepticism by the prevailing power structures. The potential profits that lay in solid waste are known, and disputes between different stakeholders involved in resource recovery have been documented. The issues need to be analysed carefully, both from the theoretical and the empirical points of view, in order to design a practical framework that takes into account the assets and barriers present in inclusive waste management.

The conceptual pillars of this chapter are as follows: a) *governance* and *deliberative democracy*, where new forms of partnerships and participatory governance initiatives are offered to address the political and social context in waste management; b) *social and solidary economy* as an overall framework focusing on collective over individual objectives and outcomes; and finally; c) *participatory resource management*, because participation is key in developing innovative solutions to our current problems. This triple theoretical approach to the issue of inclusive waste management will provide insight into the concrete experiences of co-management arrangements that are already in place in solid waste management.

Two case studies will illustrate the hands-on experiences with inclusive recycling schemes in two cities in the metropolitan region of São Paulo. The cases inform about concrete examples of waste management and possible policy outcomes. They

---

1 Part of the results were presented at the Seminar: *Impacts environnementaux et socio-économiques des options de valorisation et de recyclage des déchets solides municipaux pour les collectivités de petite et moyenne tailles* (Rabat, 1–2 June 2005).

highlight the many difficulties and hurdles that constantly have to be overcome in order to reach practicable co-management arrangements. Important additional questions are surfacing and need to be contemplated: Who owns the recyclables once they are discarded? How can we guarantee the access of economically disadvantaged groups to these resources? How can recycling become the motor for responsible consumption? In theory, co-management of waste resources seems to be a sustainable solution; in praxis, however we will see that it takes more than just a responsive government or an organized recycling initiative to work out. Participatory waste management is a complex endeavour that tackles waste management and social inclusion concomitantly.

My thesis here asserts that the participation of organized recycling groups in waste management is essential in the process of defining better ways of dealing with solid waste. The recyclers need to have a say in the framing of the policies for this sector. The activity they perform is crucial to sustainability. It reduces spending on landfill and mitigates general environmental health costs that remain externalized. The impact waste has on climate change has not yet been calculated; nor has it explicitly been discussed in the political debate on waste management, consumption and production. Once we have transparent accountability of these true costs and we understand how consumption and disposal of waste impacts on our environmental health, the economic value of resource recovery will crystallize and inclusive waste recovery policies will become more popular.

One challenge in this process is how to promote equity among the different stakeholders involved in resource recovery and prevent or mitigate eventual conflicts between them. Emerging questions on governance, co-responsibility, and power relations between the stakeholders in waste management need to be asked. Conflicts between the stakeholders' distinct perceptions of urban waste as a common property, and structural barriers such as outdated policies or overly bureaucratic and expensive formalities to legalize recycling groups, are key elements in our discussion. The literature on the topics highlighted in the previous chapters will help us get a theoretical and practical understanding, which can then be translated into sensible actions.

Governance and Deliberative Democracy

The complexity of governance is not easy to capture in a simple definition. Governance means steering societies and organizations and is mainly about decision-making and the related processes. It is difficult to define governance because of the interdisciplinary nature of the concept. Pierre and Peters (2000) provide a detailed review of the diverse perspectives on governance from the literature. The authors discuss the concept from a perspective of structures and processes. The approach taken here is best described as *governance as networks* and *governance as communities*. Networks comprise a wide variety of actors, including government and non-governmental organizations. What makes today's networks different is their concerted and cohesive organization, providing them with some power in policy making. The authors state that 'the development from government towards governance the decreasing reliance on formal-legal powers – has clearly strengthened the position of the policy networks' (Pierre and Peters 2000, 20). Governance as community sees that communities have agency to solve common problems. It is often seen as the *Third Way*, understood as a standpoint '…between state and the market' (Pierre and Peters 2000, 21) that reflects non-governmental agency and relies on ample participation.

Directing society by making and implementing policies is the central activity in governance. Different governance forms involve different levels of co-ordination and control. Governing has been defined by Kooiman (2003, 4) as 'the totality of interactions in which public as well as private actors participate, aimed at solving societal problems or creating societal opportunities; attending to the institutions as contexts for the governing interactions; and establishing a normative foundation for all those activities'. Further, 'governance can be seen as the totality of the conceptions on governing' (Kooiman 2003, 4). It is important to note that these new forms of governance allow for mutual, interactive learning – from practical problem solving to institutional learning and even to a cross-scale learning at higher levels of governance.

The term *good governance* '…is about both achieving desired results and achieving them in the right way…[which] is largely shaped by the cultural norms and values of the organization, there can be no universal template for good governance' (Institute on Governance n.d., website). The term generally characterizes those governments that have complied with certain expectations. Swilling and Hutt (2001), for example, define *good governance* as a synonym to…good administration (…*'bonne governance est synonyme de bonne administration'*), the democratisation of politics (…*'le bonne governance concerne la democratisation politique'*), the relation between the state and civil society (…*'il concerne les relations entre l'Etat et la societe civile'*), or the fact where the civil society is responsible for development (*'il designe le fait pour la societe civile d'assumer la responsabilite du developpmenent'*) (2001, 177–8). Civil society is a widely contested notion, referring to collective action based on shared interests, purposes and values. 'Civil society commonly embraces a diversity of spaces, actors and institutional forms, varying in their degree of formality, autonomy and power' (Centre for Civil Society 2004, website). Amin and Thrift propose a transformation towards '…radical democracy, where democracy requires the democratization of institutions and the empowerment of subaltern voices in a politics of vigorous but fair contest between diverse interests' (2004, 140).

Participation is a central characteristic of deliberative democracy. The notion of deliberative democracy stems from critical theory and political ecology. It envisions a different form of democracy, where inclusiveness and power redistribution are pivotal. Participation is understood as the procedure to ensure that '…the "better argument", rather than coercion or manipulation, will determine the outcome' (Zwart 2003, 24). Hence its philosophy is based on inclusiveness, reflexivity and social learning (Rosenberg 2007; Petts 2001).

Petts (2001) provides valuable insights on the application of deliberative processes in the particular context of waste management, showing the prominence of social responsibility and collective learning of the process. In this case community advisory committees and specific evaluating boards have been set up by the local authorities in order to improve the solid waste management situation. Forsyth (2005) provides another example for deliberative public-private partnerships in the context of the Philippines and India. He affirms that allowing greater public participation in the policy formulation may become an important new form of local environmental governance. Later we will see how deliberative democracy is complementary to social economy.

'Deliberating together and co-operating with others on efforts providing community-wide benefits breaks down negative stereotypes and leads to new, positive working relationships, grounded on trust' (Weber 2003, 198). The process itself helps to unify the community. Empirical results underline that '…the extensive networking means that institutions and decision-makers who used to be inaccessible to many in the community…are now only a phone call away because of the trust that networking has created' (Weber 2003, 198).

True public participation in policy-making is more than just consultation. It requires transparent democratic processes, forums for deliberation and true participation of different stakeholders. This means participants are empowered and can perceive themselves as having a real stake in the decisions to be made. In the wider literature it is recognized that citizen and community engagement is a prerequisite for more effective and more legitimate policy outcomes (Smith and Beazley 2000). Multiple stakeholder participation involves complex and often time-consuming processes. However, the final results promise higher sustainability, due to its larger acceptability. When people are part of a deliberation process, there is a sense of ownership and the agreed results have higher potential for validation and acceptability. It is also a process that builds citizenship, because it stresses inclusiveness and co-responsibility. Stratford and Jaskolski (2004) discuss a case study of deliberative democracy, applied to the construction of sustainable communities. Sustainability is increasingly associated with local participation, empowerment and deliberative democracy. Through participation, active citizenship and ecological skills can become refined and actions towards greater sustainability outcomes can be reached. The process seeks procedural change in governance to promote citizenship and participation as well as substantive changes in governance '…in adherence to the flawless exercise of communicative rationality, including faith in the capacity of deliberative democracy to engender consensus among informed equals' (Stratford and Jaskolski 2004, 321). The sustainability framework emphasizes the '…central role of active citizenship in participatory structures of governance' (Stratford and Jaskolski 2004, 312).

Participatory processes are demanding in terms of time and energy commitment. Personal barriers to participation are usually the lack of time, skills, knowledge, information, money, self-interest and political will. Furthermore, participation is not always constructive. It can also be oppositional and reactive to specific actors' or stakeholders' interests, rather than strategic and co-operative (Davidson 2003). The difficulties faced in participatory and inclusive governance and management regimes are many sided. David Harvey stresses that 'Foucault has again and again pointed out that discourses of power, attached to distinctive mediating institutions (such as the state apparatus or, more informally within the worlds of education, religion, knowledge production and the media) typically play their often overwhelming disciplinary and authoritarian role. Hegemony becomes the focus of political struggle. Imposing conceptions of the world and thereby limiting the ability to construe alternatives is always a central task of dominant institutions of power' (2001, 197). The author further discusses how the personal is always political, and how personal change is intrinsically related to social change. 'Through changing our world we change ourselves. We cannot talk, therefore, about social change without at the same time being prepared, both mentally and physically to change ourselves' (Harvey 2001, 200). That means that new governance forms also evoke necessary changes on the personal level. People in power need to become aware of the importance of sharing power in constructing more sustainable societies. Harvey also recognizes that '...in bringing persons together into patterns of social and political solidarities, there are as many traps and pitfalls as open paths to change' (2001, 201). Participation is not necessarily a guarantee for good governance. We will discuss some of these *pitfalls* further under participatory strategies, specifically from the perspective of the case studies presented later.

The Framework of Social Economy

Social economy brings social justice issues and values, such as co-operation, redistribution and reciprocity, into the economy. During the Second International Congress for the Globalization of Solidarity, hold in Quebec, in October 2001, it was agreed that '...social solidarity-based economics puts the human beings at the centre of social and economic development' (Fisher and Ponniah 2003, 91). 'Solidarity economy designates all production, distribution and consumption activities that contribute to the democratization of the economy based on citizen commitments both at a local and global level' (Fraisse, Ortiz and Boulianne 2001, 4). Like the Lima Declaration of 1997, '...it is built on a collective economic, political and social project that brings out about a new way of conducing politics and establishing human relationships on the basis of consensus and the activity of citizens' (cited in Fisher and Ponniah 2003, 91). In this declaration the focus is on contentious problems of income and wealth redistribution, social inclusion, solidarity and co-operation. Social justice is not a new concept. Moulaert and Ailenei (2005) provide a historical overview of the rise of the concept out of collective structures, from the Egyptian corporations, Greek funds for the ritual organizations of funerary ceremonies and Roman colleges of craftsmen. The first German and Anglo-Saxon *guilds* appeared in the eleventh century, the *confraternities* in France during the early *Middle Ages*. The professional or trade *corporations* and first *compagnonages* developed in France during the fourteenth century. Similar concepts of organization re-emerged during the nineteenth century in Europe (Moulaert and Ailenei 2005, 2039).

Solidarity economy happens at the local, neighbourhood and community scale, and through networking also on the global scale. This new form of doing economy creates synergies between actors (such as local authorities, private enterprises, the state, citizens) and it generates workplaces by offering new services and new ways of production, which is so important during our times where the trend is towards growing unemployment (Moulaert and Ailenei 2005).

Community-based enterprises are an important aspect of social economy. Community-based enterprise is defined as '…community acting corporately as both entrepreneur and enterprise in pursuit of the common good' (Peredo and Chrisman 2006, 310). This type of entrepreneurial arrangement is becoming increasingly popular in forestry, agriculture, small-scale industries and services. Social economy is considered an innovative form of tackling poverty and exclusion – an alternative to conventional bureaucratic and economic approaches, which have been unable to resolve or mitigate the situation of the poor. This approach involves institutional innovation, which can mean new governance arrangements and decision-making mechanisms, as well as an innovative understanding of economy as a sector, which has primarily a social purpose. Micro-credit is a tool that transfers assets to the poor so that they have control over their situation and can change it themselves. It helps farmers to farm, fishers to fish, and artisans to produce and sell their products. Local economies, particularly those in poor areas can be empowered with micro credit. The conventional banking system does not include the poor segment of society and numerous examples confirm that this has further magnified marginalization and the inability to overcome poverty. Micro-credit schemes such as the *Grameen Bank*, originated it Bangladesh, or the *Credite Rurale* in Italy have transformed the deplorable livelihood conditions of the rural poor in many regions in the world over the past decades. The scheme is being adapted to specific situations, generally benefiting minorities and the disadvantaged. Nevertheless, access to information and technical support are still rare for the large masses of urban and rural poor.

Co-operatives as well as community, neighbourhood and interest associations are essential players in social and solidarity practices (Portes and Moreira 2004). Curbing unemployment is a major target in social economy, and the main strategy is to value the local and creative workforce and to share assets and provide solidarity networks. Innovative ways of producing and doing business are arising in many locations in Brazil and in other parts of the world, particularly in developing countries. Paul Singer, a key Brazilian thinker and the current federal secretary for solidarity economy, believes that this reorientation is the right step towards humanizing work, because it values but also demands initiative and involvement from the workers' side. Important experiences are happening through these new economic and social practices, and skills are being developed through community-oriented, participatory development. Social movements that recover abandoned industries, like the Movimiento Nacional de Fábricas Recuperadas por los Trabajadores or the Movimiento Nacional de Empresas Recuperadas, emerged originally in Argentina and Uruguay during the 1980s and are now gaining momentum also in Brazil (Magnani 2003, IBASE/ANTEAG 2004). In some well-known instances, workers have successfully taken over the management and production of bankrupt enterprises, have generated new employment and have benefited the local development by recovering the local economy (Magnani 2003).

Brazil is an interesting case, because social economy advanced significantly under the previous and current federal government. Institutional changes have been implemented with the Secretaria Nacional de Economia Solidária and the Conselho Nacional de Economia Solidária to foster these initiatives. There is also the *Fórum Nacional de Economia Solidária*, an important tool to promote networking, sharing and co-operation on a national level. Several networks have been set up to co-operate in production and commercialization, so that local products reach local consumers quickly (Singer 2003). Events like the *World Social Forum, Local Agenda 21*, and other benchmarking happenings like meetings, seminars, congresses and exhibitions on the subject have also helped disseminate the idea of social economy. These growing complex regional, national and worldwide networks help promote and diffuse the ideas around the new ways of building and maintaining sustainable communities by virtue of different economic practices.

Co-governance in Resource Management

The literature describes different forms of co-governance. In forms of co-governance the essential element is that interacting parties (groups, stakeholders) have something in common to pursue. Interrelations among the parties are based on the recognition of inter-dependencies. This is often also called multi-stakeholder approach. 'This governance approach focuses on the interactions taking place between governing actors within social-political situations' (Kooiman 2003, 7). 'Co-governance in its varying appearance may be an answer, a reaction to or an expression of what [the author sees]…as a major societal development, the tendency towards growing societal interdependence and inter-penetration…Co-governance means utilizing organized forms of interactions for governing purposes' (Kooiman 2003, 97).

Collaboration and co-operation are basic principles of shared resource management. 'Collaboration is a highly diverse, volatile and complex form of co-governing arrangements. It represents in a direct way, societal diversity, dynamics and complexity in governance, and by doing so illustrates many co-governance issues' (Kooiman 2003, 99). The basic principles of these co-operative regimes are also discussed under *game theory*. 'Governing actors will co-operate under conditions involving mutual interests, limited numbers, and common concerns about the future, and will provide the necessary institutions, in the shape of self-enforcing agreements based upon principles of reciprocity' (Kooiman 2003, 100).

The various co-management modes depend on the level of co-operation different parties bring to the process and their willingness to co-operate. 'Communicative governance considers actors as reasonable citizens' and therefore a shared understanding is usually possible (Kooiman 2003, 100). Public participation in governance is in place when those involved in governing are willing to reach '…inter-subjective understanding for co-governing' (Kooiman 2003, 100). In this case the processes represent the public interests, as opposed to the usual methods of governance, which focus on the needs of only the most powerful actors. Its implementation is more time consuming, but the exchange of information leads to a greater understanding of problems and solutions, and the stakeholder interests become more apparent, improving the chances for a better understanding and even offering the hope of reaching a consensus. Here we can clearly see the importance of broad and continuous participation of stakeholder representatives in the governance process.

Another collaborative mode described by Kooiman (2003) is the *Public-Private Partnership*, a mutual interdependence between the various parties involved. Community-based enterprises are good examples of public-private partnerships. These partnerships usually occur when there is a financial-economic reason. Here, '…trust, mutual respect, adaptation, accountability are fundamental criteria' (Kooiman 2003, 102). This form of partnership is quite dynamic and follows a rapid pace of adaptation to new developments. As a pro-active measure it is therefore advisable to have coherent dispute resolution forms in place.

Co-management is defined by the sharing of responsibilities between government agencies and users or stakeholders for the well-being of the resource – for example to prevent overexploitation and to regulate fair access (Kooiman 2003). It means finding a shared understanding between government and community-initiated regulations. It is participatory rather than hierarchical; decentralized instead of centralized; and the process happens through active participation of the different parties in public policy making rather than just consultation. 'By involving the knowledge of the users in governance, results will produce more adequate governing measures' (Kooiman 2003, 103). This means that '…users involved willingly accept the regulations as appropriate and consistent with their persisting values and worldviews' (Kooiman 2003, 104).

In resource co-management, the key idea is to share responsibility. for a sustainable resource use by having them participate in all stages of the resource management process. The sharing of responsibilities between users and the state is more likely to generate long-lasting, widely accepted resolutions. Providing a forum for discussion – giving the different stakeholders a voice and an opportunity to express their views – is crucial in this process. The different standpoints are recognized and sometimes challenged in achieving a negotiated consensus. The process itself leads towards more legitimate measures and to the design of instruments that can work and have a greater compliance.

Often, government officials, who don't have the lived experience, have an inadequate understanding of the dynamics of poverty. Limited knowledge about poverty, its effects and dynamics and the prioritized needs of the poor have often resulted in inadequate developments in infrastructure and service delivery. Participatory processes can revert this through a shared learning process (Plummer 2000), which is also expressed in interactive learning, meaning learning from each other's learning. More so, what is called 'double loop learning' (Argyris 1976) has the potential to be transformative. It starts with looking at the fundamental root causes for a problem, and identifies and implements possible action for change. 'Double loop learning' requires us to question and scrutinize fundamental assumptions and values. One hurdle in exercising this type of learning is the fact that some actors may perceive such strategies as threatening the status quo. Governing actors, when challenged, may prefer to avoid shared learning and resist possible change.

Kooiman (2003) further develops the concept of co-operation into what is broadly known as networking. The information flow within networks is of a horizontal and an interdependent nature between actors. Networks can play a vital role in a new form of co-ordination: ' from hierarchical control to horizontal co-ordination' (Kooiman 2003, 105). Collaborative networks are emerging, particularly in the context of economic solidarity, where new channels and connections are created, bringing together and strengthening local and global struggles (Mance 2000). Networks have the potential to stir a silent revolution by changing opinions, attitudes and behaviours, and by evolving into collective political actions, finding ways of promoting, reinforcing and expanding on such moments in more spheres of life and struggle. The potential of these new organizational arrangements is considerable in times of political crisis of representation. The Internet connects us widely through different networks, from professional to private interests. In the less affluent world networks rely more on personal and other more traditional forms of communication. The part of society that is not able to participate is clearly disadvantaged, because they don't have the means to access and contribute to networks. Networks express societal diversity and often indicate the urge for collective action. They aggregate views, agendas and philosophies into a co-ordinated and concerted effort to act upon.

Laws and regulations are not the only elements shaping community participation in public policy framing. A balanced view of what laws can do is necessary. However, collective action on the municipal level depends on creating laws and policies that facilitate community participation. Under these conditions a constructive platform can support, simplify and accelerate the integration of participatory processes. An effective legislative framework provides greater certainty that municipal officials will really act, and act consistently. Existing laws must be questioned, principally if they only represent the needs of the powerful.

Collaborative forms of local policy shaping enhance the decentralization of processes and devolve decision-making power and responsibility to the local levels of government. As a consequence, the governing body is closer to the people, and power is returned to the local level. As this removes political and administrative control at the state level, there is the risk of resistance and political manoeuvring against participatory processes. Plummer (2000) speaks about the constraints placed on municipalities because administrative frameworks often lie outside their control. 'Much of the decision-making process is controlled by higher levels of government through the administrative apparatus, which may impose legal and regulatory obligations, may force informal requirements, and may exercise control through the financial dependence of the municipality' (Plummer 2000, 19). Ideally, higher levels of administration act as facilitators.

What affects participation in decision-making? From his experiences in participatory municipal planning Plummer (2000), emphasizes the role of skills and knowledge, employment, education and literacy, cultural beliefs and practices, gender, social and political marginalisation (inclusion), and community views over participation. As main constraints to effective municipal resource management for participation he underlines lack of trust in the municipal institutions, overall bureaucratic inertia, prevailing technical and prescriptive orientation to project development, insufficient level of training, and lack of skills to create an enabling environment. These and other aspects hinder the participation of representatives and stakeholders in resource co-management.

Active participation of the involved stakeholders is essential to adequate resource management, and to deliberative democracy. Harvey (2001) underlines that repressive and hierarchically structured forms of governance do not ensure vitality, and that for political participation, social movements are crucial to ensure the vitality (Harvey 2001). Urban social movements can also be understood as motors to ensure the strengthening of a fair and equitable government and as barometers to monitor their impact and progress. In developing countries, social movements play a critical role in setting priorities. The best solutions to poverty and injustice are those that stem directly from the actual experience of the poor, from the barriers they have to face and the institutions they lack (Soto, 1989). This means opportunities need to be made available for an ample participation and adequate representation of these interest groups.

In the metropolitan region of São Paulo many of the recyclers are organized in diverse community initiatives and co-operatives providing employment, improved working conditions, and increased environmental education. The following two case studies tell the experience from two recycling co-operatives that have been marginally involved in local policy shaping. It is very difficult for the often resource-less and disempowered recyclers to participate in collective action. Recycling co-operatives do not necessarily provide a higher income to the recyclers, particularly not during the initial phase of their organisation. The collective and solidary experience attracts members to a co-operative. In the metropolitan region of São Paulo usually more women than men are involved in the co-operatives. The reasons for a possible gender specific engagement in solidarity and co-operative arragements in the recycling sector needs to be further researched. Shared work conditions are more attractive to women also because they provide greater work flexibility, which is important given the triple work load most women face in Brazil.

Case Study: Livelihoods on the Edge

In 2005, I conducted research in Ribeirão Pires, a small city with approximately 116,000 inhabitants, at the fringe of the metropolitan region of São Paulo and entirely located within the protected watershed of Lake Billings (see Figure 6.1). In 2004, the local government supported the creation of a new recycling co-operative, *CooperPires*, in charge of door-to-door collection of recyclables in several neighbourhoods of the city (see Illustration 6.1). Initially 22 recyclers were part of this initiative and participated in the capacity building activities promoted by the municipality. They learned about quality and efficiency of selective collection and separation, health and risk protection, financial administration and group organization. As part of the support provided during this formation phase until December 2004 (*incubação*), a social worker from the municipality accompanied the weekly meetings of the group, helping them with administratative issues, conflict management and transportation logistics. The city also maintained several volunteer recycling collection points and guaranteed the transportation of the members to the recycling station.

With the change in local government at the end of 2004, official support for the programme was suspended and the municipal technicians in charge of helping the co-op were removed. This severely impacted on the co-operative. Since then the number of the co-op members has fluctuated between 18 and 24. During the peak of the political crisis due to the government change, in mid 2005, the number went down

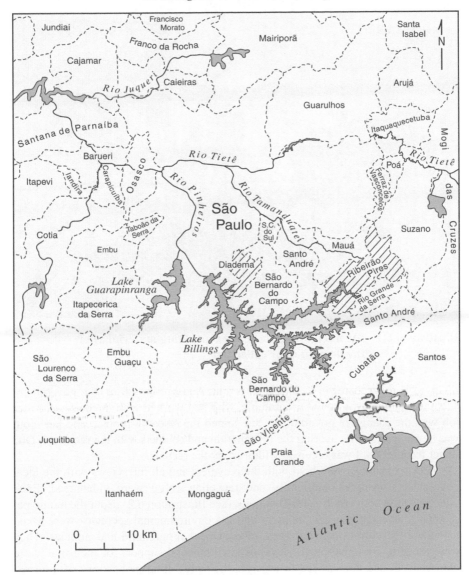

Figure 6.1   Localization of Ribeirão Pires

Sources: CPLA/SEMA 1997 and IBGE 1991.

Cartography: Ole J. Heggen.

to 12. However, half of the founding members still remained in the co-operative. The public-private partnerships set up in the past were in jeopardy, due to the dismantling of the programme. The purpose of my research was to gain the attention of the government and the public by documenting this new situation and the negative effects the change has had on the recycling co-op. The expected outcome was to re-initiate the dialogue between governmental and non-governmental stakeholders

**Illustration 6.1   Members of *CooperPires* separating material after
the door-to-door collection**

involved in this cause, via a discussion forum. Ana Maria Marins, the government
agent in charge of the social programme, who had worked with the co-op, lost her
job with the change in government. She joined the research project, and her input
was key in terms of recovering the information and contacts with the recyclers. Due
to her recognition, I was received with trust by the recyclers.

   Weekly meetings were held with the recyclers and all interviews with the local
government, business community and recycling co-op were videotaped. Each
interview took between 20 and 40 minutes (see Illustration 6.2). Both the municipal
secretary for economic development and the environmental secretary were asked
to outline the current official waste management programme and policies. They
were asked to comment on the level of governmental support for inclusive waste
management, and encouraged to describe their prevailing perceptions of informal
and organized recycling activities in the city. They were also asked whether under
their current mandate they would support inclusive, selective waste management.
The head of the local business association (ACIARP) was interviewed, as well as
three of the directors from local businesses that had established partnerships with
CooperPires. The businesses had donated material and equipment and had also
lobbied for the recyclers by voicing their support of the recycling programme.

   When we finalized the fieldwork and data collection a public seminar was held
to disseminate the findings, to gather feedback and to discuss with the different
stakeholders – including the government officials from the current and previous
administration – solutions to the catastrophic situation of abandonment of the
recyclers. Another seminar was held in March 2006, when the video was edited. We

**Illustration 6.2 Interview with Sr. José (*CooperPires*) and Fábio
from the *Fórum Recicla São Paulo***

then presented the video to the wider public and invited some guest speakers to debate the theme: inclusive waste management. Local politicians, including the mayor and the secretary for the environment and infrastructure, the business community, representatives from *CooperPires* and a representative from the recyclers' movement in São Paulo (*Fórum Recicla São Paulo*) and the general public were present.

One of the local government agents who had supported the co-operative explains that: '...due to the change in government at the beginning of this year, there was a discontinuity in the contacts with the community and the support for the co-operative. So now we have to start to regain that support, which was good, well accepted and quite positive. We have to re-stimulate and provide continuity for participation'. A member of the co-operative directly shows how the discontinuity in the public policies threatens the recyclers' livelihood: '...the salary is always going down and down, and most of the men who worked in the co-operative, who were the strongest have already left because the salary was so low...If the salary continues to go down we cannot afford to stay in the co-operative'.

As a strategy to overcome this situation, the government agent further understands that: '...in principle we need to re-establish the contacts, consult with the stakeholders, perceive and evaluate the current necessities of the co-operative in order to be able to support and train them so that they will be empowered and have autonomy. Autonomy is the key. More than recycling material, more than improving the environment, it is about recycling the hope of people'.

'Other enterprises could also participate in this programme because they could donate materials that would otherwise be discarded, and these materials could be used for other purposes, for example to build carts or shelves or desks. This would be a good way to participate in the programme'; is the opinion of the head of the business community (ACIARP). He further indicates: '...today's business needs to meet certain social and environmental standards. It's more than just about clean production. Enterprises benefit from incorporating environmental sustainability and from social equity. The business gains by providing its employees with environmental education and awareness. It also reflects on the marketing of their products'. The secretary for economic development underlines the fact that: '...the environmental cause should not be treated as a marketing exercise but rather followed up with real actions. Selective collection of waste is crucial for the city and for the world. As we are situated in a protected water catchment area, the importance and responsibility is even greater'.

The businesswoman we interviewed emphasizes: '...without participation effective recycling programmes are difficult to achieve. This depends on an educational process wherein society as a whole is involved'. Furthermore she explains: 'We entered this project because we believe that it is about recovering citizenship of people that were socially excluded. We need to have more government participation in providing education for everyone'.

The case highlights the susceptibility of new organizations, such as *CooperPires*, particularly because it involves a disenfranchised and excluded population. It also shows the detrimental impacts short-term measures and discontinuity in government programmes can have on vulnerable groups. Government participation is essential to generate a sense of co-responsibility. As highlighted earlier, participation means collective learning, which has the potential to generate the understanding of complex and often diverse positions, and which can enable a more sustainable policy design.

Recycling co-operatives are particularly vulnerable during their initial phase. Informal recyclers are marginalized by the rest of society and often suffer from lifelong exclusion. Disappointing – sometimes government induced – experiences may further reiterate preconceived ideas together with other negative life experiences. These have often made the recyclers suspicious and distrustful. Consequently many avoid collective working schemes. In some cases co-operatives are created as a result of internal leadership. However, more often they are a result of government or non-governmental initiatives. A strong long-term support and strengthening of the recycling groups is necessary, tackling different areas – from strategies that support the resolution of conflicts, address issues of health and hygiene, to building self-confidence and expanding the knowledge on commercialization and other technical aspects of the recycling process.

According to Sr. José, the leader of *CooperPires*: '...the video will help disseminate our cause...it shows our battle and it reinforces our struggle...it shows our work...The video provides us with a chance to improve our environment, to improve our situation just a little bit...because we are currently kind of down and the video has provided us with a lift in our spirit, which is very important'. He and other recyclers suggested the production of a second video to instruct new co-op members on how to efficiently collect and separate the recyclables and how to create awareness on the household level. The process of the video production has facilitated the dialogue between the co-operative and the new government. It has put pressure on the government to support the co-operative. The response from the business people was also positive in stimulating the continuation of their support. ASCIARP is playing a major role in maintaining the dialogue alive with the local government and in helping with the formalization of the co-operative.

Many bottlenecks have been identified, of which the most serious is the low economic return for co-op members. Their salaries were below the poverty rate, almost four times less than the official minimum value. The economic uncertainty of the co-op's future is a constant stress factor that directly affects the livelihood of the members. The lack of minimum sanitary infrastructure influences their health and comfort level. As of March 2007, there were still no sanitation facilities; members had to wash their hands at an open hose. Finally, the fact that *CooperPires* is not yet formalized as a co-operative restricts the organisation's ability to attract funding and support from official government programmes. The bureaucratic process is time-consuming and expensive. It involves persistence in following up on the formal requirements and it involves fees to be paid. Furthermore, the lack of formalization means restrictions regarding the access of social benefits and micro-credit, besides the negative effects from the lack of recognition of the collective effort.

Over the past year the situation has slightly improved, mainly because *CooperPires* is now part of the project *Participatory Sustainable Waste Management* (PSWM) and has received support in their negotiations and formal requests. The number of participants is once again up to 22 as of August 2007.

The success of recycling programmes depends on the efficient functioning of secondary material markets (Hornberg 1998). It is particularly important for these groups to not depend on the middlemen, because they pay significantly lower prices. The co-op participates in the PSWM initiative on collective commercialization. This has helped to increase their earnings. In March 2006, members earned Reais $130 (US$61.30) compared to Reais $50 (US$19.40) in March 2005. They now also receive a food basket each month, through the federal programme *bolsa familia*, and the municipality has approved the construction of toilets (which as already mentioned had not yet been built by early 2007) and transportation support (which has been granted since March 2007).

Active outreach involves purposeful attempts to engage and inform community members, through staging events, meetings, activities, and the distribution of written or oral information. Passive outreach involves making resources available for community use (Weber 2003). This case study portrays an active research piece, where different tools of community outreach were used in order to support the recycling co-operative in their struggle for social and economic justice.

Case Study: Innovative Public Policy in Diadema

The city of Diadema, situated next to the metropolis São Paulo (see previous Figure 6.1) has experienced rapid population growth from 70,000 people in the early 1970s to a total population of 365,000 in 2007. It is primarily an industrial town, and used to be a 'dormitory city' in the periphery of São Paulo. In the past Diadema was famous for its extremely high crime rates, visible poverty and environmental degradation. Despite the region's overall decline in industrial employment, its service sector has increased significantly over the past decade and so has the informal economy. 13 per cent of the population in Diadema lives in extreme poverty and remains illiterate.

This municipality is the first city in Brazil where organized, independent recycling groups are officially in charge of collecting, selecting and commercializing recyclables and are paid for the service. In 2004, the municipality created the recycling programme *Vida Limpa* (Clean Life). Triage facilities were set up in each of the city's five sub-catchments within the municipal boundaries. The city has supported the recycling groups in the creation of associations, so-called OSIPs (*Organização Social de Interesse Público*), to carry out the door-to-door collection and separation of recyclables. The city provided training activities to expand the capacity of the groups. Sixty locations were set up at schools, parks, government offices, and businesses to where the population could take their recyclables. Another 13 locations are planned (see Illustration 6.3).

Illustration 6.3 Door-to-door collection in Diadema

In Diadema too, there was a loss of momentum during the transition of governments in 2005, although the political party did not change. Government transition always generates destabilization and often disrupts the existing official programmes. In late 2007 there were five recycling groups, which are part of the association *Pacto*

Ambiental (Environmental Pact). The programme started with 62 recyclers. In late 2007 it involves five groups with a total of 21 women and 28 men in charge of collecting and separating recyclables. The programme received support from various government sectors, and in June 2004, a municipal law was passed institutionalizing the recycling network as part of the city's solid waste programme. The law recognizes the benefits from the partnership in waste management between the government and the recycling association. While the municipal government is supportive to the programme it still faces difficulties in addressing the demand for capacity building of informal and organized recyclers.

Diadema is the first municipality in Brazil to pay for the amount of recyclable material selectively collected under the government programme, *Vida Limpa*. On 23 December 2005, the local government signed a partnership memorandum with *Pacto Ambiental*, guaranteeing its members regular pay for the collection and diversion service. The city values this programme as a contribution to improve urban cleanliness, public health, and environmental protection, besides the benefit of generating employment (see Illustration 6.4).

Illustration 6.4   Community building through the *Vida Limpa* programme

On average, the association collects 60 tons of household recyclables every month. The amount of materials recycled through the association is growing (see Figure 6.2). As of 2006, they received Reais $38 (US$16.27) per ton of recyclable material collected; the same amount it would cost the city to deposit one ton of household waste at the landfill. In addition, the programme also collects construction waste. As in all municipalities in the metropolitan region of São Paulo, irregularly deposited construction waste is a widespread problem. The municipality of Diadema pays Reais $15.86 per ton for the preventive removal of construction waste, whereas the corrective collection would cost the city Reais $31.25 per ton.

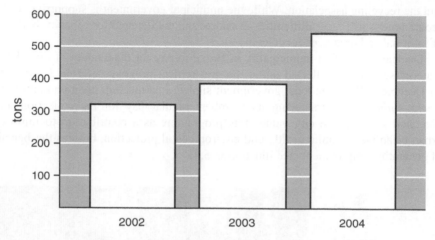

Figure 6.2   Increase in collection of recyclables in the municipality of Diadema

Source: Municipality of Diadema, *Programa Vida Limpa*. Diadema, 2007.

Cartography: Ole J. Heggen and Jutta Gutberlet.

The participants in the *Vida Limpa* programme earn approximately Reais $380 (US$162.70) per month per person, by 2007. The groups can achieve higher prices because they are able to sell directly to industry. Paper prices, for example, fluctuate between Reais $0.21 per kg for small-scale middleman, Reais $0.30 for medium-scale enterprises, and Reais $0.40 when sold directly to the industry. The programme is not yet self sufficient, and depends on the technical and financial support from the government. One of the future challenges is to expand collective commercialization and to add further value to their products. Informal recyclers are socially and economically excluded, disempowered, and generally lack environmental awareness. There is still little acknowledgement within the government and a general lack of recognition of the benefits from working with those groups instead of against them.

Innovative Governance: Participation and Empowerment

Recycling is a waste management alternative and at the same time it is a strategy to diminish unemployment. Innovative, community-based waste management systems are emerging, involving co-operatives and associations in organized door-to-door municipal programmes. Nas and Jaffe (2004) describe an interesting case of co-management in Jamaica, where the informal recycling system has been integrated into the formal waste management plan. Similarly, after recognizing the benefits of informal recycling in the waste management system in some Mexican cities, governments have started to actively support this activity with appropriate policy changes, and by encouraging the formation of co-operatives (Medina 1997, 2000). The integration of informal recycling into community-based programmes and co-operatives can provide many social, economic and health benefits to recyclers. Co-operative arrangements have more bargaining power in achieving support from the government and local community in providing education, improved living and working conditions, specific loans and scholarships, among others for informal recyclers. Medina further mentions cases where recycling co-operatives have life and accident insurance, similar to other city employees (2003).

There are other notable successful examples of recycling co-operatives in Latin America. In Colombia the *National Recycling Program* is an advanced initiative, with regional marketing associations and social development programmes for its members. Amongst other initiatives, the *Associacion Cooperativa de Recicladores de Bogotá* has recently created the housing co-op *Pedro León Trabuchi* to relocate the recyclers, members of this association, that were living under hazardous conditions in Bogotá. In Mexico progress has been made in terms of legalizing the informal and organized recycling activities, encouraging the formation of co-operatives and micro-enterprises as well as awarding concessions in waste management to organized recycling groups (Medina 2003). Medina (2003) further underlines that the 'policies supporting *carretoneros*' [recyclers'] activities are humane, environmentally sound, socially desirable, and economically viable' (9). Here the local government has a unique opportunity to develop organized door-to-door selective waste collection arrangements as a strategy to tackle the social and environmental agenda. It takes an integrated perspective and a sense of social responsibility to understand that embracing waste management from an inclusive perspective builds stronger communities and enhances community health.

A pre-requisite for such a paradigm shift is the willingness to accept and support innovative socio-environmental approaches in waste management. Technical and engineering solutions can only capture certain facets of the picture that usually exclude the social perspectives. Hence these conventional answers can only partially solve the problems. Although still widespread and perpetuated by our governments, landfills and incineration are not acceptable solid waste management solutions. Valuable resources are wasted and new environmental, social and economic predicaments are created. Inclusive waste management provides the opportunity to generate employment and to redistribute income. As an organized collective the recyclers have a larger bargaining power to receive better prices. With better incomes there is the chance for the collective to improve the working conditions and to cultivate personal growth (Medina 2000). Through shared work recyclers have a greater ability to meet industrial needs for clean, baled, crushed and sorted material.

Co-operatives and associations can commercialize collectively and do not need to sell to middlemen, who are known for exploiting the autonomous recyclers. Hence prices are better and the standard of living of the recyclers can improve. Furthermore, as co-operatives or associations, the recyclers have better chances to access micro-credit and to gain support for infrastructure through partnerships with the government, business or community. Logistics and strategies for collective commercialization as well as access to micro-financing are current and contentious topics in Brazil. Some of the strongest co-operatives are already organized in secondary regional networks (federations or second degree co-operatives), collaborating in the commercialization and sharing their assets and information. For smaller and less well structured groups, bureaucratic hurdles involved in the legalization of co-operatives or associations remain the major impediment to this development.

The case of *CooperPires* has shown the difficulties related to political instability and weak institutions. The fragmented nature of government is a frequent hindrance for efficient and economic resource management. Lack of communication between agencies, excessively complex structures, overly extensive bureaucratic procedures, prone to rivalry, corruption and inefficiencies are often major institutional barriers to implementing inclusive policies. Structural changes and fundamental mentality shifts are necessary. Methods to effectively communicate and reiterate key messages that challenge the existing social, economic, and political conditions need to be in place to also achieve increased environmental health. Under collaborative and participatory governance structures, information is considered a public right rather than a privileged commodity (Weber 2003, 248).

The *CooperPires* example confirms the importance of public-private partnerships. Here the local stakeholders' involvement was pivotal in exercising pressure on the government to be co-responsible for the recycling programme initiated with *CooperPires*. Strengthening the institutional, regulatory and human capacity is an important pre-requisite for programmes like these to flourish. Collective approaches increase their bargaining power and help recyclers access more lucrative markets. Recycling associations and co-operatives empower their members and generate a group identity, which can significantly increase social cohesion and hence be translated into social capital. Social development work with recycling co-operatives can contribute to the strengthening of the members' identity and awareness and it can help build their self-esteem.

Non-governmental organizations play an important role in organizing informal recycling efforts, particularly in capacity building. These programmes can help provide a better standard of living for its members. They dignify the recycling occupation, and are able to provide important links between the government and the community (Baud et al. 2001; Nas and Jaffe 2004). Beall points towards the role non-governmental organizations play; '…[they] have been recognized as important urban partners, reaching where governments and international agencies cannot reach… effectively addressing urban poverty and representing the urban poor' (2000, 850).

Among the biggest threats for community-based approaches in waste management is the increasing trend towards privatization of this sector, with a dominance of large-scale enterprises and multinational corporations controlling this sector (Fahmi 2005; Magera 2003; Ogu 2000; Lee 1997). These private actors prioritize incineration and landfill as waste management techniques. More recently, they have begun to play a central role in the recycling business. These developments have serious effects on the informal recycling sector, as described by Fahmi (2004), Kaseva and Mbuligwe (2005) and Post et al. (2003). The experiences show that small-scale initiatives usually are unable to compete with large corporations from the private sector. Unless a different paradigm is in place, whereby the social and environmental service recyclers provide is recognized, and unless recyclers are involved in resource recovery and are paid adequately for their work, they will remain in poverty and uneven development will be perpetuated. Privatization of waste management, particularly selective collection and recycling activities in developing countries, poses serious threats to the livelihoods of the urban poor (Fahmi 2005). It is short sighted and unsustainable to follow this market-oriented trend without looking at the wider social implications.

The first case reveals the political and economic barriers to inclusive waste management in the city of Ribeirão Pires. Local governments often undervalue existing social assets, such as the recycling co-operative. Here the situation was far from being participatory and a top-down decision had excluded recyclers from an income-generating activity. Political decisions have seriously affected the livelihoods of the members from the local recycling co-operative. This example highlights the difficulties and prevailing constraints in making inclusive and participatory public policies in waste management. It underlines the quest for ample recognition of the social and environmental benefits of this activity.

The second case discussed the innovative public policy in Diadema that has opened new horizons for resource recovery as a poverty reduction measure and as a strategy to improve environmental health. Here informal selective collection and recycling has provided a concrete chance to improve the livelihoods of impoverished and excluded. Integrated solid waste management based on partnerships between the government, recycling groups and local business as described in this case is a viable option and has now already been adopted in several other cities in Brazil and in other countries. Ahmed and Ali (2004), Baud et al. (2001), McBean et al. (2005), Medina (2000), Sarkar (2003) and other scholars have recognized the significance of such programmes in improving the livelihoods of recyclers and in providing opportunities for better informed and more sustainable policy initiatives. Of course there is room for further improvements to be made. The association in charge of the selective collection in Diadema, for example, is still very fragile. All aspects touching the members' livelihoods need to be enhanced – from occupational health to access to formal education. Group assets, such as trust, participation and confidence, need to be fortified. Empowerment of the recyclers reflects in greater political saying when it comes to defining waste management policies and environmental health issues. This might also raise concerns, particularly among the political class that feels frightened by the power from below. There are concrete examples where politics has hindered the recyclers' organization because of the fear from the local government over the political influence of the recyclers' movement (Gutberlet 2008 in press).

Some of the pivotal lessons learned from both experiences are:

1. Inclusive recycling programmes need the support from the government. They need to be integrated into the municipal solid waste programme and should not be treated as a separate programme.

2. To increase efficiency, these recycling programmes need to be decentralized and adapted to the prevailing local geographic conditions. This also means giving autonomy to each group involved, while adopting an appropriate overall working framework.

3. Planning on the watershed basis seems to work very well. More so because it takes the topography into consideration, which is decisive for pushcart-driven waste collection.

4. The relationship between recycling groups and the municipality needs to be based on professional relations. Paternalistic structures and approaches are not constructive, but rather maintain the processes that create dependency.

5. A social assistance approach needs to focus on the empowerment of these groups and on strengthening their autonomy.

6. Recovering the dignity and citizenship of recyclers needs to become a public responsibility. Overall, there are many social, environmental, and economic gains for the municipality from the collection and separation of recyclables; this needs to be fully recognized and valued.

The prevailing view of waste and waste recovery has to change. There needs to be a shift from treating our waste carelessly as useless material towards recognizing it as a resource and contributing to its recovery. Instead of recyclers being perceived as a nuisance and treated with aversion, they have to be recognized as environmental service providers. Empowering informal recyclers through capacity building, information and participation enables social inclusion.

Participatory waste management is an effective anti-poverty strategy. Organized recycling programmes provide an opportunity to enhance public environmental awareness with the recyclers performing the role of environmental agents that assist in achieving a better quality of waste separation at the source (Bolaane and Ali 2005). This has become evident in an experience established long ago in Cairo, where 60,000 recyclers, locally called the *Zabaleen* (rubbish collectors) collect and sort recyclables from the 16 million inhabitants. Although their activity is widely recognized they now face ferocious competion with multinational waste management companies (Fahmi 2005).

Inclusive waste management also contributes to strengthening democratic processes. As Young puts it: 'Instrumentally, participatory processes are the best way for citizens to ensure that their own needs and interests will be voiced and will not be dominated by other interests' (1990, 92). Participatory waste management is an approach to guarantee more durable decision-making and contributes to the construction of more sustainable communities. Participatory resource management (co-management), involving informal and organized recyclers as stakeholders is an effective way of tackling crucial challenges in waste management. The analysis of international experiences provides valuable information about the assets and constraints of such arrangements. It is up to citizens to demand from our governments a more appropriate waste management. These experiences need to be disseminated so that cities and countries can learn from each other.

Tenório and Rozemberg (1997) refer to social participation and citizenship building as processes empowering individuals to exercise their rights to democratically construct their own destiny. Social participation and citizenship building can be implemented through collective organization of participants: for example, by opening spaces for discussion within and outside of the community to define priorities, elaborate action strategies and create channels to engage in the dialogue with the government. Participatory budgeting, although not the ultimate solution, is an approach that is well confirmed in its practical value of building active citizenship and is more than just political discourse and talk. It is important that the government opens up planning and managing structures for society to participate, so that new democratic processes of co-management can be created (Fleury 2003). Selective collection and recycling are sensible and concrete forms where Governments can demonstrate actions rather than only discourses of sustainable development.

References

Ahmed, S. and Ali, M. (2004), 'Partnerships for Solid Waste Management in Developing Countries: Linking Theories to Realities', *Habitat International* 28, 467–79.

Amin, A. and Thrift, N. (2002), *Cities: Re-imagining the Urban* (Cambridge: Polity Press).

Argyris, C. (1976), *Increasing Leadership Effectiveness* (New York: Wiley).

Baud, I., Grafakos, S., Hordijk, M. and Post, J. (2001), 'Quality of Life and Alliances in Solid Waste Management', *Cities* 18(1), 3–12.

Beall, J. (2000), 'From the culture of poverty to inclusive cities: Re-framing urban policy and politics', *Journal of International Development* 12, 843–56.

Bolaane, B. and Ali, M. (2005), 'Organized recycling in Gaborone, Botswana', *Engineering Sustainability* 158(54), 223–34.

Centre for Civil Society (2004), 'What is civil society? London School of Economics', [website] http://www.lse.ac.uk/collections/CCS/what_is_civil_society.htm, accessed 26 November 2007.

Davidson, J. (2003), 'Citizenship and sustainability in dependent island communities: the case of the Huon Valley region in southern Tasmania', *Local Environment* 8(5), 527–40.

Fahmi, W.S. (2004), 'Urban sustainability and poverty alleviation initiatives in garbage collectors community: A stakeholder analysis of the Muqattam "*Zabaleen*" settlement in Cairo', Cambridge: *ENHR Conference* (2, 6 July 2004).

—— (2005), 'The impact of privatization of solid waste management on the Zabaleen garbage collectors of Cairo', *Environment and Urbanization* 17(2), 155–70.

Fisher, W.F. and Ponniah, T. (2003), *Another World is Possible: Popular Alternatives to Globalization at the World Social Forum* (London: Zed Books).

Fleury, S. (2003), 'Políticas sociais e democratização do poder local', in Vergara, S.C. and Correa, V.L. de A. (Orgs.), *Propostas para uma Gestão Pública Municipal Efetiva* (Rio de Janeiro: Editora FGV).

Forsyth, T. (2005), 'Building deliberative public-private partnerships for waste management in Asia', *Geoforum* 36, 429–39.

Fraisse, L., Ortiz, H. and Boulianne, M. (2001), *Solidarity Economy: Proposal Paper for the XXI Century.* Workshop on Solidarity Socio-Economy, [website] http://ecosol.socioeco.org/documents/81pdf_fnl15en.pdf, accessed 26 November 2007.

Gutberlet, J. (2008), 'International university partnership: reflections and learning experiences from Brazil and Canada', in Boothroyd, P. and Dharamsi, S., *Universities and Participatory Development: Lessons from International Experience* (in press).

Harvey, D. (2001), *Spaces of Capital: Towards a Critical Geography* (Edinburgh: Edinburgh University Press).

Hornborg, A. (1998), 'Towards an ecological theory of unequal exchange: Articulating world system theory and ecological economics', *Ecological Economics* 25, 127–36.

IBASE/ANTEAG (2004), *Autogestão em Avaliação* (São Paulo: ANTEAG (Associação Nacional dos Trabalhadores em Empresas de Autogestão e Participação Acionária)).

Institute on Governance (IOG) (n.d.), Ottawa, [website] http://www.iog.ca/, accessed 26 November 2007.

Kaseva, M.E. and Mbuligwe, S.E. (2005), 'Appraisal of solid waste collection following private sector involvement in Dar Es Salaam city, Tanzania', *Habitat International* 29, 353–66.

Kooiman, J. (2003), *Governing as Governance* (London: Sage Publications).

Lee, Y.S.F. (1997), 'The privatization of solid waste infrastructure and services in Asia', *Third World Planning Review* 19(2), 139–61.

Leonard, M. (2000), 'Coping strategies in developed and developing societies: The workings of the informal economy', *Journal of International Development* 12, 1069–85.

Magera, M. (2003), *Os Empresarios do Lixo. Um Paradoxo da Modernidade* (Campinas: Editora Atomo).

Magnani, E. (2003), 'El cambio silencioso Empresas recuperadas en la Argentina', [website] http://www.estebanmagnani.com.ar/wp-content/elcambiosilencioso.pdf, accessed 14 September 2007.

Mance, E. (2000), *A Revolução das Redes – a Colaboração Solidária Como uma Alternativa Pós-capitalista à Globalização Atual* (Petrópolis: Editorial Vozes).

McBean, E.A., del Rosso, E. and Rovers, F.A. (2005), 'Improvements in Financing for Sustainability in Solid Waste Management', *Resources, Conservation and Recycling* 43, 391–401.

Medina, M. (1997), 'Informal recycling and collection of solid wastes in developing countries: Issues and opportunities', The United Nations University: Institute of Advanced Studies, *Working Paper* 24.

—— (2000), 'Scavenger cooperatives in Asia and Latin America', *Resource Conservation and Recycling* 31, 51–69.

—— (2003), 'Serving the unserved: Informal refuse collection in Mexican cities', *Collaborative Working Group on Solid Waste Management*, Dar Es Salaam: Workshop on solid waste collection that benefits the urban poor (9–14 March 2003).

Moulaert, F. and Ailenei, O. (2005), 'Social Economy, Third Sector and Solidarity Relations: A Conceptual Synthesis from History to Present', *Urban Studies* 42(11), 2037–53.

Moulaert, F. and Nussbaumer, J. (2005), 'Defining the Social Economy and its Governance at the Neighbourhood Level: A Methodological Reflection', *Urban Studies* 42(11), 2071–88.

Nas, P.J.M. and Jaffe, R. (2004), 'Informal Waste Management: Shifting the Focus form Problem to Potential', *Environment, Development and Sustainability* 6, 337–53.

Ogu, V.I. (2000), 'Private sector participation and municipal waste management in Benin City, Nigeria', *Environment and Urbanization* 12(2), 103–17.

Peredo, A.M. and Chrisman, J.J. (2006), 'Toward a theory of community-based enterprise', *Academy of Management Review* 31(2), 309–28.

Petts, J. (2001), 'Evaluating the Effectiveness of Deliberative Processes: Waste Management Case-studies', *Journal of Environmental Planning and Management* 44(2), 207–26.

Pierre, J. and Peters, B.G. (2000), *Governance, Politics and the State* (New York: St. Martin's Press).

Plummer, J. (2000), *Municipalities and Community Participation. A Sourcebook for Capacity Building* (London: Earthscan).

Portes, L. and Moreira, M. (2004), 'Cooperativas geram trabalho e renda; Maioria dos cooperados são mulheres', *Diaro Oficial* 8(453), 3–5.

Post, J., Broekema, J. and Obirih-Opareh, N. (2003), 'Trial and error in privatization: Experiences in urban solid waste collection in Accra (Ghana) and Hyderabad (India)', *Urban Studies* 40(4), 835–52.

Rosenberg, S.W. (2007), 'Rethinking Democratic Deliberation: The Limits and Potential of Citizen Participation', *Polity* 39(3), 335–60.

Sarkar, P. (2003), 'Solid Waste Management in Delhi – A Social Vulnerability Study', *Proceedings on the Third International Conference on Environment and Health*, Chennai, India.

Singer, P. (2003), 'As grandes questões do trabalho no Brasil e a economia solidária', *Proposta* 30(97), 12–16.

Smith, M. and Beazley, M. (2000), 'Progressive regimes, partnerships and the involvement of local communities: a framework for evaluation', *Public Administration* 78(4), 855–78.

Soto, de H. (1989), *The Other Path: The Invisible Revolution in the Third World* (New York: Harper & Row).

Stratford, E. and Jaskolski, M. (2004), 'In pursuit of sustainability? Challenges for deliberative democracy in a Tasmanian local government', *Environment and Planning B: Planning and Design* 31, 311–24.

Swilling, M. and Hutt, D. (2001), 'Johannesburg, Afrique du Sud', in Onibokun, A.G., *La Gestion des Déchets Urbains, des Solutions pour L'Afrique* (Paris, Karthala, Ottawa: CRDI).

Tenório, F.G. and Rozenberg, J.E. (1997), 'Gestão Pública e Cidadania: metodologias participativas em ação', *Cadernos Gestão Pública e Cidadania* 7.

Weber, E.P. (2003), *Bringing Society Back In: Grassroots Ecosystem Management, Accountability and Sustainable Communities* (London: MIT Press).

Young, I.M. (1990), *Justice and the Politics of Difference* (Princeton: Princeton University Press).

Zwart, I. (2003), 'A Greener Alternative? Deliberative Democracy Meets Local Government', *Environmental Politics* 12(2), 23–48.

Chapter 7

Conclusion

Beyond Recycling: Social Development and Environmental Benefits

With the diffusion of *free market*, *trade liberalization* and other market policies in operation (such as structural adjustments led by the *Word Bank* and the *International Monetary Fund*) under the *Washington Consensus*, the dominant economic development model of the *Global North* is being rapidly applied to the *Global South*. Most developing countries are troubled by natural resource depletion and environmental contamination as well as all the social impacts related to unemployment, exclusion and loss of cultural heritage. The change in the development paradigm towards primarily economic determinants has increased inequities and has created disparities in the quality of life, with scandalous differences in opportunities and outcomes between poor and rich.

In many developing countries the prevailing economic system is based on structures that have created and are maintaining inequity and unequal development (Sen 1999). Young (1994) states that oppression underlines the condition of exclusion. She defines five *faces of oppression* as being the faces of social injustice. As to Young (1994), the following aspects relate to oppression:

1. *Exploitation*: wage, focus on the institutional structures: family, workplace;

2. *Marginalization*: as racially marked groups, material deprivation, exclusion from citizenship;

3. *Powerlessness*: as inhibition of the development of one's capacities, lack of decision-making power, exposure to disrespectful treatment because of the status one occupies;

4. *Cultural imperialism*: universalization of dominant group's culture and experience, defined from the outside, influenced by images, stereotypes;

5. *Violence*: physical violence through attacks on the person or property.

Among the main attributes that widen the social and economic gap is the widespread practice of corruption, taking advantage of privileges (also termed *neo-colonialism*) and *power over* relations. The generation of economic wealth is often based on the exploitation of others and the environment. Social exclusion and environmental degradation are common outcomes of this development. Here we are talking about more than just economic poverty. Social exclusion consists of the separation of individuals or groups from the rest of society because of economic deprivation or/ and social and cultural segregation. The issue needs to be addressed at the level of processes, institutions and individuals. The face of exclusion in the Global South is distinct and follows certain patterns that characterize the condition of large sectors of the population:

1. *Educational divide* with difficulties for many (particularly women) to complete formal education, resulting in lack of reading and writing abilities.

2. *Digital divide* due to the disadvantage in accessing written and digital information, in not having access to computers and the internet.

3. *Informality* because of less opportunities in formal workforce.

4. *Economic exclusion* as a consequence of unemployment, underemployment and exploitive wage levels.

5. *Health risks* because of insecure living conditions (precarious basic infrastructure and unreliable public services) as well as unhealthy work situations.

6. *Social exclusion* visible in the lack of leisure and recreational infrastructure; negligent, paternalistic and populist policy making, reinforcing the status quo; and disempowerment.

Exclusion and precariousness are growing in our communities. At the same time as our consumption-oriented society creates new and multiple necessities related to consumer goods, it makes access to them impossible for a significant part of the population. The above listed *faces of social exclusion* are widespread among the urban poor, including the dwellers of squatter settlements in the metropolitan region of São Paulo, places that are repeated in other cities and countries in the developing world. These topics addressed with social exclusion are high ranking on the political agenda for social change. Concomitantly, new social movements are emerging to address some of these social predicaments. Informal and organized recyclers are key players in the struggle for social justice; they address the concerns raised here through resource recovery and recycling.

Participation for Social Transformation

Social participation means providing individuals with the right to democratically construct their present and future, which implies empowering particularly those who are socially excluded (Tenório 1997). To put this into practice involves collective organization of the participants. It means enabling them to enter discussion spaces within and outside the community boundaries, and involves defining priorities and action strategies. Spaces and opportunities for public discussions need to be created in order to generate participation and the possibility for dialogue. The challenges are manifold and directly related to the predisposition of local governments to provide for these spaces and to include the different voices. Fleury (2003) points out the potential for social innovation, by way of transforming governmental structures and policies permitting the inclusion of citizens who were excluded before.

In his book *Bringing Society Back*, Weber (2003) describes the advantages of collaborative arrangements, where people have equal access to power, information and action, resulting in a shared generation of outcomes. Collaborative arrangements inspire and create 'increased opportunities for innovation and successful problem solving' (Weber 2003, 70). The quality and the level of participation, of course, matter significantly. Weber (2003) talks about *participant norms*, which are: a) *inclusiveness* (all have the right to participate), b) *civility and respect* for others, (education and awareness) c) *integrity and honesty* in communication and action (non violent communication), d) *dual role* as community member and representative of particular interests (understanding for individuality and collectivity), e) commitment to a '*balanced*' *approach* (such as balance between local knowledge and scientific knowledge) and finally, f) *trust* as obligation.

> In considering social action, specifically that devoted to some form of change, [we] are faced with the need to recognize how people's diverse circumstances and experience nevertheless result in sufficient stimulus for specific purposes and identities to form, through which power relations are engaged and actions are chosen and implemented. Thus, to act – to participate, contribute, challenge or resist – involves intersecting social processes, where difference, identity and power are negotiated. (Panelli 2004, 191)

Participation not only has the potential to transform processes and institutions; it also transforms individuals. Leaders, stakeholder participants or community members are changed through the process of collectively promoting change. The process empowers individuals and supports personal growth. Since participation is mostly a volunteer activity, it is also a process that demands time and, often, financial resources. The wider benefits of inclusive processes need to be recognized and opportunities for exercising participation need to be created by the government and society at large. 'The construction of local solidarities and the definition of local collectivities and affinities is a crucial means, whereby the person becomes broadly political' (Harvey 2001, 203).

Participatory approaches recognize the knowledge and experience of the local population. Increasingly, academics acknowledge the importance of local knowledge in rural, natural resource management settings. The trans-generational wisdom can contribute significantly in finding corrective measures that address today's apparent environmental predicaments deriving from mechanized resource extraction and use as well as so-called modern development practices that put the economic profits before long-term sustainability. The lessons learned through traditional local people's knowledge also need to be recognized in the context of urban resource management. Local voices reveal community assets and hindrances; understanding these issues is particularly important in the process of formulating consistent policy frameworks. Bottom-up, participatory approaches are driven by practical outcomes. They constitute a paradigm shift in epistemological understanding, which traditionally focuses primarily on representational knowledge.

Park (1999) points towards the question: What is valuable knowledge? Here another issue emerges, related to the notion of expertise. In the context of co-management Weber defines knowledge as expanding '...beyond scientific, bureaucratic and organized interest expertise to include technical expertise in the community and citizen generalities with a community perspective' (2003, 72).

Citizens' input on social capital, such as community life, local social assets or other values prevalent in the community contain an equally important level of information, which has to be valued. *Place based stakeholders* (Weber 2003) often have critical contributions to make. Participatory processes have a high potential to empower participants. Those who are never heard, who are excluded and discriminated against can be empowered by voicing their concerns and actively contributing to the generation of new knowledge. Participatory research is widely used where communities face problems with social relations, lack of basic infrastructure and services, or political disenfranchisement (Park 1999). It brings community members and other stakeholders together to deal with the problem, resulting in collaborative research procedures, adapted to a specific situation.

The rural, natural resource context described for rapid rural appraisals also applies to participatory research in urban settings. 'Participatory research is the process of empowering the community to re-search its bio-physical and socio-cultural environment to generate new knowledge and understanding which will serve as bases for the formulation of strategy, resource management and livelihood and building confidence in sustaining its efforts towards community-based [...] resource management' (Ferrer and Nozawa 1998, 7). Pain (2004), furthermore, acknowledges that participatory research is a quest for more relevant, morally aware and non-hierarchal practices. Action research begins with the identification of the problem and the stakeholders (Thiollent 1997). It involves interviewing and

Illustration 7.1   Participatory research in Diadema

group discussions in the definition of action plans for change (Annison 2002, Hay 2001). A decisive outcome of this approach is the empowerment of disenfranchised populations, which have historically been *outside* or excluded from the traditional research process: be it of generating wealth and formal knowledge. Action research recognizes the political power of the participating population and perceives them as agents of transformation of unjust social and economic relations. Participation throughout the research process allows participants to achieve a sense of ownership, and to consequently become integral to the process of improving the social conditions (Park 1999) (see Illustration 7.1). Raco (2000) points out that these approaches, such as participatory budgeting, are increasingly recognized as central to policy making.

A New Waste Recovery Paradigm

The preceding chapters have touched on the questions: Who has to pay for the costs of *managing* our waste? How do we treat our waste? Waste management can employ many people; however, with increased level of automation fewer jobs are available in this sector. In countries with a large informal economy, the number of people involved in the collection of recyclables is substantial and waste is an obvious resource to them. Who benefits from the business of waste management? Who gets involved in the recovery of the resources? Can informal recycling contribute to diminishing the waste of resources? Is there a chance for redistributing income?

There is a close relationship between social and environmental justice issues and waste. Social justice concerns the degree to which a society maintains and supports the institutional conditions necessary for '...developing and exercising one's capacities and expressing one's experience and [further] participating in determining one's action and the conditions of one's action' (Young 1994, 37). It is a condition that requires high levels of respect, democratic values, and participation.

The distributive paradigm defines social justice as the morally proper distribution of social benefits and burdens among society's members. It is about the distribution of wealth, income, material resources, rights, opportunity and power. 'Justice equally requires, however, participation in public discussion and processes of democratic decision-making. All persons should have the right and opportunity to participate in the deliberation and decision-making of the institutions to which their actions contribute or which directly affect their actions' (Young 1994, 91). She underlines the importance of the intrinsic value of participatory democratic processes.

Earlier I discussed the role of participation under deliberative democracy and the construction of more sustainable communities. Co-management regimes based on stakeholder participation are concrete forms that can guarantee the participation of different voices in defining the access and use of resources reaching more accepted resolutions. Inspiring literature is emerging about integrated resource management, particularly experiences dealing with water resources. Chapter 6 discussed the transfer of this management knowledge to the domain of solid waste resources. The informal and organized recycling sector is still mostly excluded from official waste management programmes despite the large number of people working in informal recycling. The activity is not recognized as a means of resource recovery. For the public at large, waste is considered an end of the line and is seen to be of no use, just something that needs to be discarded and made invisible. The benefits for environmental health and global sustainability remain unrecognized. Everywhere the people who perform this informal activity are often discriminated by the public and police, although collaborative and respectful partnerships between recyclers and individuals or business can happen.

Large corporations usually have a vested interest in exploring markets related to waste. Waste management collection and deposit operations are paid by tonnage of waste collected and deposited or incinerated. As a consequence these enterprises are not concerned with waste reduction or recycling. By granting long-term contracts to these firms, municipalities often lock themselves into unsustainable waste management schemes. Sometimes cities sign contracts for more than 20 years, as did the city of São Paulo.

Informal recycling is a centuries old activity. Over time, the scope and level of dependency of the people involved in this activity has changed. Today even women, children and elderly people are involved. Child labour is unacceptable and needs to be combated with adequate access to education and guaranteed minimum salary. Recycling has always been a livelihood strategy for the most impoverished and excluded in the absence of a social security system. Informal recycling is likely to increase in periods of economic crisis with high unemployment and poverty. It is an adaptive response to the scarcity of financial resources (Ali 1999). Although I am not in accordance with all of Soto's writing, he highlights that often '…informality [is] defined as the refuge of individuals who find that the cost of abiding by existing laws in the pursuit of legitimate economic objectives exceeds the benefits' (1989, xxii). Prejudices against these people need to be corrected. 'Informal activities are not carried out in a chaotic or anarchic world, as many have assumed [….] informals have clear and specific interests and a previously unsuspected level of organization governed by rules that they have spontaneously developed to replace those which the state has failed to provide' (Soto 1989, xxii).

The social networks and ease of entry into the informal economy have been stressed as being a major advantage for the recyclers, generating positive effects on employment opportunities and the distribution of income. Research has noted the benefits of engaging in informal employment for the individual, notably the direct cash income, and flexibility in hours (Gerxhani 2004; Snyder 2004).

Informal recyclers everywhere experience strong prejudices; they are often humiliated, and are seldom treated with respect when they collect material in the street. The public generally associates them with dirt and disease, or they are perceived as nuisances and sometimes as criminals. There is often dispute between carts and cars in the street and the public habitually does not see the valuable service provided by these people. Successful inclusion of this sector into municipal selective collection programmes is highly dependent on a collaborative relationship between the recyclers and the wider community.

Organized recycling co-operatives can render many services to cities, including street sweeping, public and private cleaning services, and other local initiatives besides door-to door selective collection of recyclables and environmental education. Particularly in low-income neighbourhoods without basic public services, the contribution to the urban environment by collecting what would otherwise contaminate roads, parks, rivers, lakes, beaches and other public spaces is significant.

Although the organization of informal recycling into co-operatives and community-based programmes has gained considerable attention in the literature, the prevailing attitude of local governments is still to exclude this activity from the recovery process. Authorities view this sector with suspicion and often refuse to admit its important role in resource recovery operations. Governments see the recyclers as a social problem instead of seeing the solution to many social and environmental problems. More so, is it important to highlight successful government programmes, such as the *Programa Vida Limpa* in Diadema, discussed in Chapter 6. Here recycling associations are responsible for selective household collection in various water catchments of the city. Associations are paid for the quantity they collect in the same way the municipality would pay when depositing the waste on the landfill. The recyclers yet need to be paid for the environmental service they are providing.

While the *modern market democracy* is often proposed as solution to widespread poverty and informality in developing countries, I believe that a paradigm shift based on the values of co-operation, participation and solidarity as defined in solidarity economy (Gaiger 2005, Moulaert and Ailenei 2005) are the key elements in the transformation of poverty, injustice, environmental degradation and resource depletion. I challenge the understanding that sees the promotion of economic growth through market institutions as the main development purpose. There are concrete examples in Brazil where solidarity and social economy are generating income, redistributing wealth, creating social capital, improving liveability and contributing to the aim of building more sustainable communities (Singer 2003). The fundamental values of reciprocity, solidarity and co-operation as well as sustainability – rather than infinite consumption and economic growth – are fostered through such an approach. There is growing recognition that a global revolution is underway, a revolution, which is not headed by a specific political party but is mobilized by social movements connected through collaborative networks. Diverse popular and participatory initiatives are integrated under solidarity and social economy, an emerging and encompassing approach in a possible post-capitalist era. A new paradigm shift embodying social and environmental values through the praxis of solidarity economy is emerging out of the breaking eggshell of the still prevailing development model.

As highlighted before, informal recycling is not only a characteristic of poor countries. In the United States and Canada, many of the excluded and particularly the homeless find a living in informal recycling. A significant number of the *binners*, as they call themselves locally, are immigrants or long time unemployed; many have physical or mental health problems or are drug addicts. The flexible and often variable source of income is a survival strategy that individuals with no or very little income depend on. There is a valuable societal contribution done by the recyclers. More could be done to support and facilitate these activities. Canadians dispose of approximately 1 billion aluminium cans per year, adding 16,000 tons to the landfills. Were these materials recycled, the equivalent of one million gigajoules of energy could have been saved or enough to heat 10,870 homes for one year (Morawski 2007).

Recycling Citizenship

Recycling is not the ultimate environmental solution. It does not tackle the waste problem at its roots, since it does not prevent the drive towards resource intensive consumption nor the production of resource intensive consumer goods. Recycling can be an opportunity for generating and redistributing income. It is an activity that already involves the most excluded. Organized recycling embodies the possibility for recovering citizenship. We need solutions that are suited to the local context: preserving natural resources and contributing to social development, which enhances the quality of life of the local community and which means generating needed employment. Progressive public policies in waste management need to focus on:

1. *Inclusion*: formatting inclusive waste management programmes with organized recycling groups (co-operatives, associations, community groups) and facilitating their articulation.

2. *Equity*: guaranteeing fair pay and social benefits for the service of resource recovery and assuring gender equity.

3. *Eco-health*: addressing all levels of health, from protecting the health of the workers to improving environmental health.

4. *Eco-efficiency*: introducing best practice in source minimisation at all production levels, aiming towards zero packaging and product waste, introducing co-responsibility for producers and consumers and intensifying resource recovery.

5. *Sustainability*: assuring that the root causes of our unsustainable and unjust production and consumption models are addressed on a long-term perspective.

Learning from local experiences and finding appropriate technologies are important components of the process to overcome the many immediate challenges and needs. The various stakeholders from within the community need to be involved in the building of awareness and construction of a vision based on the aspects outlined above. One way of addressing these challenges is to implement environmental education programmes addressing consumers and producers in waste reduction and waste recovery. Results are long-lasting when the people involved can embrace the cause and actively participate in the transformation course. Only participatory processes that address the issues of power and politics can truly address the complex challenges of changing the existing paradigm.

Active participation of local stakeholders needs to be a given in policy decision-making. This means that local authorities need to create spaces for inter-institutional and multi-stakeholder dialogue. The powerful organization of the recyclers into a national movement in Brazil is a sign for the current change that is challenging the existing paradigm (see Illustrations 7.2 and 7.3). In other Latin American countries similar processes are happening. There are more than 1,000 recycling co-ops in South America (150 alone in Colombia). The 1st World Conference of Informal Recyclers, held in Bogotá from 29 February to 4 March is an example for inclusive globulization. Here recyclers and government and non-government supporters from 43 countries discussed the assets and barriers of inclusive waste management as well as the challenges with professionalization of informal recyclers into resource recoverers. Processes like these empower the participants and bear the strong potential to create a more just and healthy society. Inclusive mechanisms are able to better address local poverty and economic disparities with specific policies directed towards redistribution and away from unsustainable production and consumption patterns. The sophisticated advertising techniques of the media, with increased power of seduction to encourage greater consumption of goods and services, are also responsible for increased consumption and disposal. To explain increased consumption, particularly in the better-off parts of the word Flávia Soares and Nelson Diehl (2006) link the concept of *modernisation of poverty* (Illich 1978) 'to an impoverishment of the mental and operational capacity to face daily problems, which ends up transformed into a need to consume' (Soares and Diehl 2006, 8).

**Illustration 7.2   Second International Congress of the National
Recyclers' Movement in São Leopoldo, 2005**

The implementation of resolutions that have emerged out of the local context is always a big challenge for the prevailing top-down power structures. Despite the obvious potential for conflicts between the different stakeholders, bottom-up is a convincing approach towards the aspired social ideals of a different development paradigm. The success stories in local development need to be communicated and the tool of video documentation can contribute to publicize these experiences. Videotaping enhances the transparency and sustainability of the process in documenting relations and outcomes.

Illustration 7.3   CooperFenix, one of the members of the national recyclers'
movement, MNCR

There is still a significant gap between theory and practice on sustainable community
development and participatory solid waste management. Despite the few advances
and the many new challenges and hurdles, there are some examples that demonstrate
interesting results of collective action to improve livelihood issues. The case of the
neighbourhood association *Pedra sobre Pedra* provides insights to the struggle on
the neighbourhood level. It shows facets of the reality of community mobilization,
the difficulties and achievements in improving their livelihoods. Hickey and Mohan
re-affirm the empowering potential of participation, being a '…catalyst to underpin
genuine processes of transformation' (Hickey and Mohan 2005, 257). The existence
of poverty is inconsistent with the goal of creating a better society. Poverty needs
to be eradicated for the construction of a new civilization. Inclusive waste recovery
is a realistic practice that tackles this request. Privatizing waste management and
passing the sector over to multinational companies is a threat for inclusive waste
management (Fahmi 2005) that needs to be reverted with sensible public policies
and collaborative public private partnerships (Ahmed and Ali 2004).

Not only a different production (Santos 2006), but also a different world is
possible, a world with the following characteristics:

- *Enhanced human security*: by stimulating the generation of employment (with labour intense practices, re-use and recycling), by eliminating health risks with the introduction of appropriate technology, and by adopting political accountability and participatory, integrated decision-making (such as: participatory budgeting, co-management).
- *Preserving environmental sustainability*: by promoting resource-conserving lifestyles and sustainable values (such as bioregionalism, slow-food, local production, organic food), through stimulating reduction, re-use, and recycling (voluntary simplicity, eco-tax, certification), and by supporting the development of reusable and biodegradable products. Drastically cutting over-consumption and moving towards responsible consumption needs to be the overarching aim.

First attempts to shift the prevailing economic paradigm have already been made. Individual thriving experiences need to be adapted to the prevailing local conditions and the information related to the practices of failure and success need to be widely disseminated. A different world is possible and a different economy is needed. The proliferation of solidarity economy initiatives worldwide underlines this claim. In Brazil alone circa 1.2 million workers are, integrally or partially, involved in solidarity economy and 1,250 social enterprises have emerged in the last five years (Mance 2007). I agree with Euclides André Mance when he says, '...if for many it is only a utopia, an ever-receding horizon of hope, for millions of others solidarity economy is a way of working, producing, commercialising, consuming and exchanging values. It is a way of satisfying individual and personal needs in the interest of the welfare of all. It is the material base of the network revolution' (Mance 2007, 8).

Organized selective collection, based on autonomy and solidarity, is a viable entry point for the excluded into a dignified life with fair livelihood conditions. Not only do we have an opportunity to tackle social and environmental problems with this activity but we also have an obligation to revert the picture of wasting resources, lives and environments. With capacity building in safe work practices, administration skills, co-operativism and environmental education the recyclers will be able to perform all facets of the environmental service related to resource recovery, from collecting and separating recyclables to educating the population about how to separate the material most efficiently at home. Over time this service can be adjusted to new emerging social and economic constellations related to consumption, recycling and reuse, so that if one day recycling is not needed because more sustainable ways of resource minimization have been found, then the category of the recyclers will have adjusted to new social and environmental service functions.

References

Ahmed, S.A. and Ali, M. (2004), 'Partnerships for solid waste management in developing countries: linking theories to realities', *Habitat International* 28, 467–79.

Ali, M. (1999), 'The informal sector: What is it worth?', *Waterlines* 17(3), 10–12.

Annison, J. (2002), 'Action research: Reviewing the implementation of a distance-learning degree programme: Utilizing communication and information technologies', *Innovations in Education and Teaching International* 39(2), 95–106.

Ferrer, E.M. and Nozawa, C.M.C. (1998), 'Community-Based Coastal Resources Management in the Philippines: Key Concepts, Methods and Lessons Learned', IDRC, [website] http://idrinfo.idrc.ca/archive/corpdocs/120639/15-120639-NOZAWA-PAPER.pdf, accessed 8 June 2007.

Fleury, S. (2003), 'Políticas sociais e democratização do poder local', in Vergara, S.C. and Correa, V.L. de A. (Orgs.), *Propostas para uma Gestão Pública Municipal Efetiva* (Rio de Janeiro: Editora da Fundação Getúlio Vargas).

Gaiger, L. (Org.) (2004), *Sentidos e Experiências da Economia Solidária no Brasil*, (Porto Alegre: Editora da UFRGS).

Gerxhani, K. (2004), 'The informal sector in developed and less developed countries: A literature survey', *Public Choice* 120, 267–300.

Harvey, D. (2001), *Spaces of Capital: Towards a Critical Geography* (Edinburgh: Edinburgh University Press).

Hay, I. (2001), *Qualitative Research Methods in Human Geography* (Oxford: Oxford University Press).

Hickey, S. and Mohan, G. (2005), 'Relocating participation within radical politics of development', *Development and Change* 26(2), 237–62.

Illich, I. (1978), *Toward a History of Needs* (New York: Pantheon Books).

Mance, E.A. (2007), 'Solidarity economics', *Turbulance: Ideas for Movement*, [website] http://www.turbulence.org.uk/solidarityeconom.html, accessed 1 September. 2007.

Morawski, C. (2007), 'Garbage in, garbage out', *Solid Waste and Recycling* 4(1), 8–13.

Moulaert, F. and Ailenei, O. (2005), 'Social Economy, Third Sector and Solidarity Relations: A Conceptual Synthesis from History to Present', *Urban Studies*, 42(11), 2037–53.

Pain, R. and Francis, P. (2003), 'Reflections on participatory research', *Area* 35:1, 46–54.

Panelli, R. (2004), *Social Geographies* (London: Sage Publications).

Park, P. (1999), 'People, knowledge, and change in participatory research', *Management Learning* 30(2), 141–57.

Raco, M. (2000), 'Assessing community participation in local economic development – Lessons for the new urban policy', *Political Geography* 19, 573–99.

Santos, B. de S. (ed.) (2006), *Another Production is Possible. Beyond the Capitalist Canon* (London: Verso).

Sen, A. (1999), *Development as Freedom* (Oxford: Oxford University Press).

Singer, P. (2003), 'As grandes questões do trabalho no Brasil e a economia solidária', *Proposta* 30(97), 12–16.

Snyder, K.A. (2004), 'Routes to the informal economy in New York's east village: Crisis, economics, and identity', *Sociological Perspectives* 47(2), 215–40.

Soares, F. and Diehl, N. (2000), *Ethical Consumption*, proposal papers for the XXIst Century – Charles Leopold Mayer Editions, used for the debate in the Alliance for a Responsible, Plural and United World, [website] http://www.alliance21.org, accessed 6 July 2007.

Soto, de H. (1989), *The Other Path: The Invisible Revolution in the Third World* (New York: Harper & Row).

Tenório, F.G. (1997), 'Gestão pública e cidadania: metodologias participativas em ação', *Revista de Administração Pública* 31(4), 101–25.

Thiollent, M. (1997), *Pesquisa-ação nas Organizações* (São Paulo: Atlas).

Weber, E.P. (2003), *Bringing Society Back In: Grassroots Ecosystem Management, Accountability and Sustainable Communities* (London: MIT Press).

Young, I.M. (1994), *Justice and the Politics of Difference* (Princeton: Princeton University Press).

Index